Ben Stacy Jerrik (Ed.)

Great Fransham

Ben Stacy Jerrik (Ed.)

Great Fransham

Norfolk, King's Lynn, Great Yarmouth, Central London

Part Press

Imprint

Permission is granted to copy, distribute and/or modify this document under the terms of the GNU Free Documentation License, Version 1.2 or any later version published by the Free Software Foundation; with no Invariant Sections, with the Front-Cover Texts, and with the Back- Cover Texts. A copy of the license is included in the section entitled "GNU Free Documentation License".

All parts of this book are extracted from Wikipedia, the free encyclopedia (www.wikipedia.org).

You can get detailed informations about the authors of this collection of articles at the end of this book. The editors (Ed.) of this book are no authors. They have not modified or extended the original texts.

Pictures published in this book can be under different licences than the GNU Free Documentation License. You can get detailed informations about the authors and licences of pictures at the end of this book.

The content of this book was generated collaboratively by volunteers. Please be advised that nothing found here has necessarily been reviewed by people with the expertise required to provide you with complete, accurate or reliable information. Some information in this book maybe misleading or wrong. The Publisher does not guarantee the validity of the information found here. If you need specific advice (f.e. in fields of medical, legal, financial, or risk management questions) please contact a professional who is licensed or knowledgeable in that area.

Any brand names and product names mentioned in this book are subject to trademark, brand or patent protection and are trademarks or registered trademarks of their respective holders. The use of brand names, product names, common names, trade names, product descriptions etc. even without a particular marking in this works is in no way to be construed to mean that such names may be regarded as unrestricted in respect of trademark and brand protection legislation and could thus be used by anyone.

Cover image: www.ingimage.com
Concerning the licence of the cover image please contact ingimage.

Publisher:
Part Press is a trademark of
International Book Market Service Ltd., 17 Rue Meldrum, Beau Bassin, 1713-01 Mauritius
Email: info@bookmarketservice.com
Website: www.bookmarketservice.com

Published in 2012

Printed in: U.S.A., U.K., Germany. This book was not produced in Mauritius.

ISBN: 978-613-5-61831-0

Contents

Great_Fransham

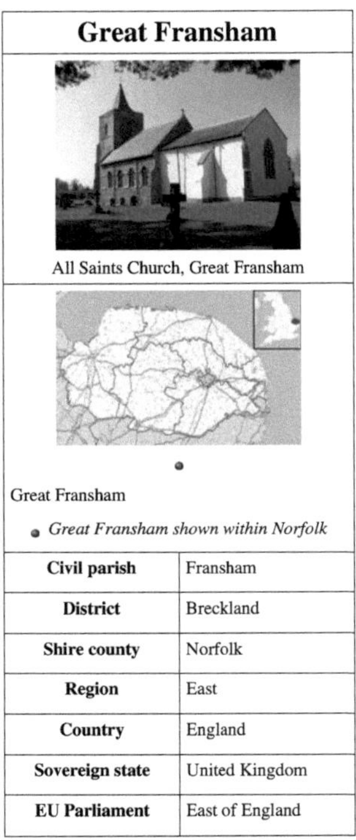

Great Fransham

All Saints Church, Great Fransham

Great Fransham

Great Fransham shown within Norfolk

Civil parish	Fransham
District	Breckland
Shire county	Norfolk
Region	East
Country	England
Sovereign state	United Kingdom
EU Parliament	East of England

Great Fransham is a village in Norfolk roughly about an equal distance between Swaffham and Dereham. There is also a Little Fransham; the two villages, both a part of the civil parish of Fransham, were once served by Fransham railway station.

All Saints Church

The church of All Saints is flint in the Early English style, consisting of chancel, nave, north porch and a square tower with spire; it contains two ancient brasses: it was restored in 1878 at a cost of about £700. The register dates from the year 1558. The living is a rectory, yearly value £552, with residence and 62 acres of glebe, in the gift of Magdalen College, Cambridge, and held since 1853 by the Rev. Vincent Raven M.A. late fellow, tutor, and president of that college. Alongside is a schoolhouse, built in 1871. Kelly's directory Cambridgeshire, Norfolk & Suffolk, in 1883 claims *"The school is for 80 children, average attendance 51, supported by voluntary rate & school pence; Miss Esther Quartermain, mistress"*.

William Crane

In 1865, at the age of 24, William Crane set up a Blacksmith's shop in the village of Great Fransham in Norfolk. Little could have William known that the name of Crane would go on to become one of the biggest names in transport in the 20th Century. William and his three brothers all followed in their Father by becoming apprenticed Blacksmiths. William went on to develop a new type of horse rake, cart wheels and farm Wagons. Later with his two sons, the business expanded further and was registered in the 1883 White's directory as "William Crane - Agricultural implement maker, joiner and builder, smith and wheel wright and church bell hanger". Due to a bad dept, they received a load of timber which provided the raw material for the production of various carts and wagons.

William died in 1906 and his two sons carried on the business which continued to expand and in 1913 Cranes acquired the former Mallons agricultural works at South green in Dereham. Cranes by this time had such an excellent reputation for its products that is secured a substantial order for gun carriage wheels and field ambulances for the army. In 1920, Edward Crane returned to the Fransham workshop to continue his Timber business and the Dereham facility branched out into road trailers in 1920. Cranes had an excellent reputation for specialist trailers and secured some significant military contracts on the back of their expertise and quality of build.

Crane's began making trucks using designs by the American-based Fruehauf Corporation in the early 1960s, with the North Walsham factory opening in 1962 and Toftwood (Dereham) in 1968. The Crane name then became intrinsically linked with Fruehauf and the Crane Freuhauf brand became synonymous with the trailer units plying their trade all over Europe, following the growth of roll on - roll off routes to the continent. Taken over by General Trailers in 1997, the company later reinstated under the Fruehauf brand. When the Crane Fruehauf factory in Dereham collapsed in 2005, laying off its remaining 345 workers, an important chapter in Norfolk's industrial history came to an abrupt end.

Railway

The Lynn & Dereham Railway was given the Royal Assent on 21 July 1845,opened in stages between 1846 and 1848 and later became part of the Great Eastern Railway. Hunt's Directory of East Norfolk 1850 shows Edgar Skeit as a 'railway clerk'. However White's History, Gazetteer, and Directory, of Norfolk 1854,lists Edgar Skeet as being station master at Great Fransham. He would spend over 30 years in this role.In the early days 4 passenger and 1 goods train would pass through the station each way daily.Great Fransham was a halt between the two major junctions of East Dereham and Swaffham.The station also had a level crossing.

The original intention of the company had been to extend their line to Great Yarmouth, via Norwich, but this plan was blocked by the rival Wymondham to Dereham scheme proposed by the Norfolk Railway. The line was closed to passenger and freight services by the Eastern Region of British Railways on Saturday 7 September 1968.

Sir Vincent Raven

Vincent Litchfield Raven was born on 3 December 1859 the son of a clergyman at Great Fransham Rectory, Norfolk and educated at Aldenham School in Hertfordshire. In 1877 he began his career with the North Eastern Railway as a pupil of the then Locomotive Superintendent, Edward Fletcher. By 1893 he had achieved the post of Assistant Mechanical Engineer to Wilson Worsdell who was then the Locomotive Superintendent. In this post he was involved for the first time with an electrification project, as the N.E.R. was electrifying the North Tyneside suburban route in 1904. This was a third rail system at 600 volts DC..In 1910, Raven becomes Chief Mechanical Engineer for the NER. He then starts the first phase of his new locomotives using both steam and electric propulsion. By 1912 Raven has begun to add superheating to many NER steam locomotives. Raven's designs continued NER traditions but he was also innovative. He oversaw the electrification of the Shildon to Newport freight line, planned further electrification, and developed a widely used cab signalling system. During World War I, Raven was Superintendent at the Royal Arsenal, Woolwich where he organised munitions production, for which he received a knighthood in 1917.] The Grouping of the railways in 1923 gave the Chief Mechanical Engineer's post to Nigel Gresley of the Great Northern Railway and Raven became a Technical Adviser. He resigned in 1924 and was appointed to the Royal Commission on New South Wales Government Railways, in company with Sir Sam Fay. He died on 14 February 1934 after heart trouble whilst on holiday with Lady Raven in Felixstowe.

Church Farm

With the arrival of the railway the Rev. Vincent Raven invested in creating a model farm to provide milk and farm produce for the rectory and to sell at Dereham and Kings Lynn markets. A new slate roofed farmhouse and pantiled group of model farm buildings were constructed on glebe land near the new railway line.

Gt. Fransham Mill

Gt. Fransham towermill was a four storey mill that stood at Mill Farm described as a very wide-towered Mill, not very high, but thick & heavily built. The mill used four patent sails to power two pairs of French burr stones. A pulley wheel was set onto the outside of the mill to allow auxiliary power via a belt from a steam engine. In 1886, two windmills were advertised to be let along with a bake office on the same site. The Mill had a history of tragedy. The Norfolk News of 10 February 1866 reported On Thursday in last week a distressing accident accompanied by a fatal termination occurred in Mr. Perkins' mill, the unfortunate victim being a respectable & steady married man in his service, named Crispin Howard. About ten minutes before eleven in the forenoon Mr. Perkins left the deceased following his usual avocation. On his return shortly after eleven, Mr. Perkins thought the mill was not working so smoothly as before he left & on looking upwards he discovered the lifeless body of his workman resting on the beam to which he had been elevated after having been drawn in & passed between the two cog wheels. From the nature of the wounds inflicted, it was obvious death must have been instantaneous. The dead man who was fifty five years of age leaves a widow & three children to mourn his loss & for whom much sympathy is felt. At the inquest held on Saturday, the only verdict that could be found in the circumstances was returned – that of "Accidental death". Just a few years later The Kings Lynn Advertiser reported in April 1874 that Jonathan Perkins, miller of Gt Fransham was fatally injured on being thrown out of a pony cart when it hit a gatepost. By the early part of the 20th century the end had come for Great Fransham Mill. The Dereham & Fakenham Times - 28 August 1909 reported the sale of the mill by auction at the George Hotel, East Dereham, on 23 August 1909. Mr. Heyhoe next offered the substantial built brick tower windmill situate at the Mill Farm, Great Fransham. This contained patent sails, two pairs of stones, shafting gear and fittings in good order and were sold subject to being removed from the occupation by 11 October next, from instructions from the trustees of Court 1246 A. O. F. (Swaffham). Mr. Crane was the purchaser for £7. This was the Crane family who owned the wagon works at Gt Fransham.

The brick tower has been long demolished however Mill Farm House, which dates from the second half of the 16th century, still exists.

Norfolk

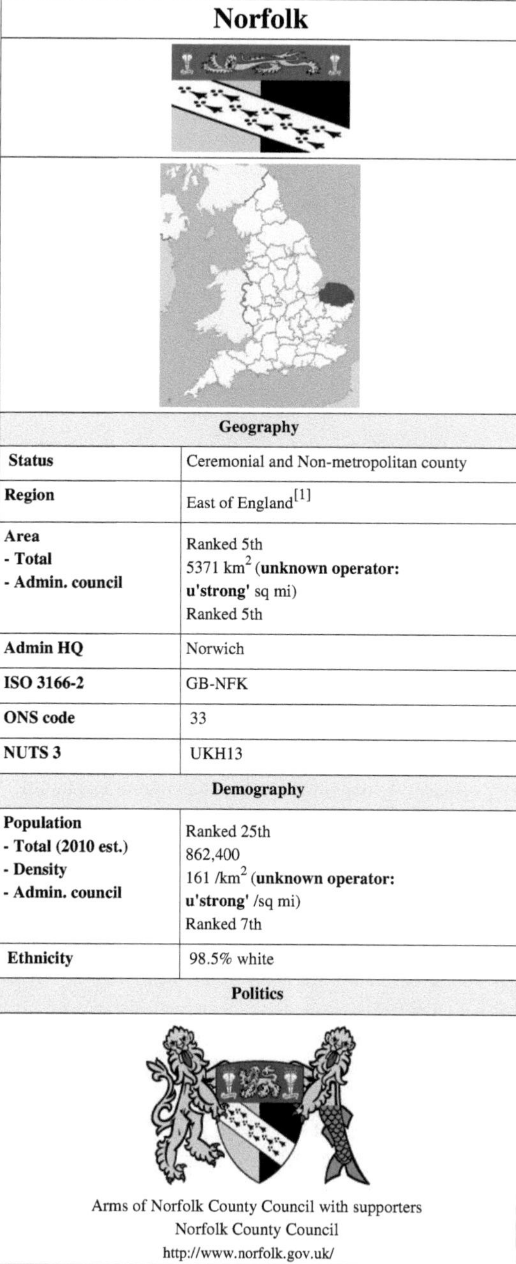

<table>
<tr><th colspan="2">Norfolk</th></tr>
<tr><td colspan="2" align="center">Geography</td></tr>
<tr><td>Status</td><td>Ceremonial and Non-metropolitan county</td></tr>
<tr><td>Region</td><td>East of England[1]</td></tr>
<tr><td>Area
- Total
- Admin. council</td><td>Ranked 5th
5371 km^2 (unknown operator: u'strong' sq mi)
Ranked 5th</td></tr>
<tr><td>Admin HQ</td><td>Norwich</td></tr>
<tr><td>ISO 3166-2</td><td>GB-NFK</td></tr>
<tr><td>ONS code</td><td>33</td></tr>
<tr><td>NUTS 3</td><td>UKH13</td></tr>
<tr><td colspan="2" align="center">Demography</td></tr>
<tr><td>Population
- Total (2010 est.)
- Density
- Admin. council</td><td>Ranked 25th
862,400
161 /km^2 (unknown operator: u'strong' /sq mi)
Ranked 7th</td></tr>
<tr><td>Ethnicity</td><td>98.5% white</td></tr>
<tr><td colspan="2" align="center">Politics</td></tr>
<tr><td colspan="2" align="center">Arms of Norfolk County Council with supporters
Norfolk County Council
http://www.norfolk.gov.uk/</td></tr>
</table>

Executive	Conservative
Members of Parliament	• Richard Bacon (C) • Henry Bellingham (C) • Elizabeth Truss (C) • Chloe Smith (C) • Norman Lamb (LD) • Brandon Lewis (C) • George Freeman (C) • Keith Simpson (C) • Simon Wright (LD)
Districts	

1.	Norwich
2.	South Norfolk
3.	Great Yarmouth
4.	Broadland
5.	North Norfolk
6.	King's Lynn and West Norfolk
7.	Breckland

Norfolk (◀ /'nɔrfək/) is a low-lying county in the East of England. It has borders with Lincolnshire to the west, Cambridgeshire to the west and southwest and Suffolk to the south. Its northern and eastern boundaries are the North Sea coast and to the north-west the county is bordered by The Wash. The city of Norwich is the county town at Norfolk which is fifth largest ceremonial county in England, with an area of 5,371 km² (2,074 sq mi).

Of the 34 non-metropolitan English counties, Norfolk is the seventh most populous, with a population of 850,800 (mid 2008). However, as a largely rural county it has a low population density, 155 people per square kilometre (or 401 per square mile). Norfolk has about one-thirtieth the population density of central London, the tenth lowest density county in the country, with 38% of the county's population living in the three major built up areas of Norwich (259,100), Great Yarmouth (71,700) and King's Lynn (43,100).[2] The Broads, a well known network of rivers and lakes, is located towards the county's east coast, bordering Suffolk. The area has the status of a National Park and is protected by the Broads Authority.[3] Historical sites, such as those in the centre of Norwich, also contribute to tourism.

History

Norfolk was settled in pre-Roman times, with camps along the higher land in the west where flints could be quarried.[4] A Brythonic tribe, the Iceni, inhabited the county from the first century BC, to the end of the first century (AD). The Iceni revolted against the Roman invasion in 47 AD, and again in 60 AD led by Boudica. The crushing of the second rebellion opened the county to the Romans. During the Roman era in Norfolk roads and ports were constructed throughout the county and farming took place.

Situated on the east coast, Norfolk was vulnerable to invasions from Scandinavia and northern Europe, and forts were built to defend against the Angles and Saxons. By the 5th century the Angles, after whom East Anglia and England itself are named, had established control of the region and later became the "north folk" and the "south folk", hence, "Norfolk" and "Suffolk". Norfolk, and several adjacent areas, became the kingdom of East Anglia, later merging with Mercia and then Wessex. The influence of the Early English settlers can be seen in the many place names ending in "-ton", and "-ham", as well as "-by" or "-thorpe" for Danish place names, In the 9th century the region again came under attack, this time from Danes who killed the king, Edmund the Martyr. In the centuries before the Norman Conquest the wetlands of the east of the county began to be converted to farmland, and settlements grew in these areas. Migration into East Anglia must have been high, as by the time of the Conquest and Domesday Book survey, it was one of the most densely populated parts of the British Isles. During the high and late Middle Ages the county developed arable agriculture and woollen industries. Norfolk's prosperity at that time is evident from the county's large number of mediaeval churches: of an original total of over one thousand, 659 survive, more than in the whole of the rest of Great Britain.[5] The economy was in decline by the time of the Black Death, which dramatically reduced the population in 1349; suffice to say that the current population has yet to equal the population before that time. Over one-third of the population of Norwich died during a plague epidemic in 1579.[6] By the 16th century Norwich had grown to become the second largest city in England, but in 1665 the Great Plague again killed around one third of the population.[7] During the English Civil War Norfolk was largely Parliamentarian. The economy and agriculture of the region declined somewhat, and during the industrial revolution Norfolk developed little industry except in Norwich and was a late addition to the railway network.

In the 20th century the county developed a role in aviation. The first development in airfields came with the First World War; there was then a massive expansion during the Second World War with the growth of the Royal Air Force and the influx of the American USAAF 8th Air Force which operated from many Norfolk Airfields. During the Second World War agriculture rapidly intensified, and has remained very intensive since, with the establishment of large fields for cereal and oil seed rape growing. Norfolk's low-lying land and easily eroded cliffs, many of which are chalk and clay, make it vulnerable to the sea, the most recent major event being the North Sea flood of 1953.

The low-lying section of coast between Kelling and Lowestoft Ness in Suffolk is currently managed by the Environment Agency to protect the Broads from sea flooding. Management policy for the North Norfolk coastline is described in the North Norfolk Shoreline Management Plan which was published in 2006 but has yet to be accepted by the local authorities.[8] The Shoreline Management Plan states that the stretch of coast will be protected for at least another 50 years, but that in the face of sea level rise and post-glacial lowering of land levels in the South East, there is an urgent need for further research to inform future management decisions, including the possibility that the sea defences may have to be realigned to a more sustainable position. Natural England have contributed some research into the impacts on the environment of various realignment options. The draft report of their research was leaked to the press, who created great anxiety by reporting that Natural England plan to abandon a large section of the Norfolk Broads, villages and farmland to the sea to save the rest of the Norfolk coastline from the impact of climate change.[9]

Economy and industry

In 1998 Norfolk had a Gross Domestic Product of £9,319 million, making it 1.5% of England's economy and 1.25% of the United Kingdom's economy. The GDP per head was £11,825, compared to £13,635 for East Anglia, £12,845 for England and £12,438 for the United Kingdom. In 1999–2000 the county had an unemployment rate of 5.6%, compared to 5.8% for England and 6.0% for the UK.[10]

Important business sectors include tourism, energy (oil, gas and renewables), advanced engineering and manufacturing and food and farming.

Wells-next-the-Sea.

Much of Norfolk's fairly flat and fertile land has been drained for use as arable land. The principal arable crops are sugar beet, wheat, barley (for brewing) and oil seed rape. Over 20% of employment in the county is in the agricultural and food industries.[11]

Well-known companies in Norfolk are Aviva (formerly Norwich Union), Colman's (part of Unilever) and Bernard Matthews Farms. The Construction Industry Training Board is based on the former airfield of RAF Bircham Newton. The BBC East region is centred on Norwich, although it covers an area as far west as Milton Keynes.

New Anglia [12] Local Enterprise Partnership has been recently established by business leaders to help grow jobs across Norfolk and Suffolk. They have secured an enterprise zone to help grow businesses in the energy sector and established the two counties as being a centre for growing services and products for the green economy.

To help local industry in Norwich, Norfolk, the local council offered a wireless internet service but this has now been withdrawn following the end of the funding period.[13]

River Wensum, Norwich.

Education

Primary and secondary education

Norfolk has a completely comprehensive state education system, with secondary school age from 11 to 16 or in some schools with sixth forms, 18 years old. In many of the rural areas, there is no nearby sixth form and so Sixth form colleges are found in larger towns. There are twelve independent, or private schools, including Gresham's School in Holt in the north of the county, Thetford Grammar School in Thetford--Britain's fourth oldest school, Langley School in Loddon, the

Norwich Cathedral: Spire and south transept.

original Norwich boys school, Norwich School and Norwich High School for Girls in the city of Norwich itself. The Kings Lynn district has the largest school population. Norfolk is also home to Wymondham College, the UK's largest remaining state boarding school.

Tertiary education

The University of East Anglia is located on the outskirts of Norwich and Norwich University College of the Arts (previously Norwich School of Art and Design) is based in seven buildings in and around St George's Street in the city centre, next to the River Wensum.

The City College Norwich and the College of West Anglia are colleges covering Norwich and Kings Lynn as well as Norfolk as a whole. Easton College, 7 miles (**unknown operator: u'strong'** km) west of Norwich, provides agriculture-based courses for the county, parts of Suffolk and nationally.

University Campus Suffolk also run higher education courses in Norfolk, from multiple locations including Great Yarmouth College.[14]

Politics

Norfolk is a shire county, under the control of Norfolk County Council. This is divided into seven local government districts, Breckland District, Broadland District, Great Yarmouth Borough, King's Lynn and West Norfolk Borough, North Norfolk District, Norwich City and South Norfolk. As of 2011 the Conservatives control all of the six districts outside Norwich, while Norwich remains under No Overall Control with Labour as the largest party.

The county is traditionally a stronghold for the Conservatives, who have always won at least 50% of Norfolk's constituencies since 1979. The countryside is, as expected, mostly solid Conservative territory,

Norwich Roman Catholic Cathedral.

with a few areas being strong for the Liberal Democrats. North Norfolk is competitive between the Conservatives and Liberal Democrats. From 1995 to 2007, South Norfolk was run by the Liberal Democrats, however the district switched back to the Conservatives in a landslide in 2007.

Norfolk's urban areas are more mixed, although Norwich and central parts of Great Yarmouth and King's Lynn are strong for the Labour Party UK. This said, Labour's dominance in Norwich has recently been stemmed by the Green Party, who are now the official opposition on Norwich City Council and also hold the greatest number of divisions within the city on Norfolk County Council.

In October 2006, the Department for Communities and Local Government produced a Local Government White Paper inviting councils to submit proposals for unitary restructuring. Norwich submitted its proposal in January 2007, which was rejected in December 2007, as it did not meet all the rigorous criteria for acceptance. In February 2008, the Boundary Committee for England, (from 1 April 2010 incorporated in the Local Government Boundary Commission for England), was then asked to consider alternative proposals for the whole or part of Norfolk, including whether Norwich should become a unitary authority, separate from Norfolk County Council. In December 2009, the Boundary Committee recommended a single unitary authority covering all of Norfolk, including Norwich.[15] [16] [17] [18]

However, on 10 February 2010, it was announced that, contrary to the December 2009 recommendation of the Boundary Committee, Norwich would be given separate unitary status.[19] The proposed change was strongly resisted, principally by Norfolk County Council and the Conservative opposition in Parliament.[20] Reacting to the announcement, Norfolk County Council issued a statement that it would seek leave to challenge the decision in the courts.[21] A letter was leaked to the local media, in which the Permanent Secretary for the Department for Communities and Local Government noted that the decision did not meet all the criteria and that the risk of it "being successfully challenged in judicial review proceedings is very high".[22] The Shadow Local Government and Planning Minister, Bob Neill, stated that should the Conservative Party win the 2010 general election, they would reverse the decision.[23]

Following the 2010 general election, Eric Pickles was appointed Secretary of State for Communities and Local Government on 12 May 2010 in a Conservative–Liberal Democrat coalition government. According to press reports, he instructed his department to take urgent steps to reverse the decision and maintain the status quo in line with the Conservative Party manifesto.[24] [25] However, the unitary plans were supported by the Liberal Democrat group on the city council, and by Simon Wright, LibDem MP for Norwich South, who intended to lobby the party leadership to allow the changes to go ahead.[26]

The Local Government Act 2010 to reverse the unitary decision for Norwich (and Exeter and Suffolk) received Royal Assent on 16 December 2010. The disputed award of unitary status had meanwhile been referred to the High Court, and on 21 June 2010 the court (Mr. Justice Ouseley, judge) ruled it unlawful, and revoked it. The city has therefore failed to attain permanent unitary status, and the previous 2-tier arrangement of County and District

Councils (with Norwich City Council counted among the latter) remains the status quo.[27]

Norfolk County Council is Conservative-controlled and led by Derrick Murphy. There are 63 Conservative councillors, 11 Liberal Democrat councillors, 5 Green Party councillors, 3 Labour councillors and 1 UKIP councillor.[28] There was a 63% turnout at the most recent local election.

Following the May 2010 General Election, Norfolk is represented in the House of Commons by seven Conservative Members of Parliament and two Liberal Democrats. The Labour Party have lost the urban areas of Norwich and Great Yarmouth in recent elections, leaving them with no Commons representative in East Anglia; the former Home Secretary Charles Clarke being a high level casualty in the 2010 election.

Norfolk Election Results

Parliamentary 6 May 2010					County Council 4 June 2009				
Party	Votes	Votes %	Seats	Seats %	Party	Votes	Votes %	Seats	Seats %
Conservative	188,944	43.1%	7	77.8%	Conservative	115,396	45.9%	60	71.4%
Liberal Democrat	121,710	27.8%	2	22.2%	Liberal Democrat	56,998	22.7%	13	15.5%
Labour	83,088	19.0%	0	0%	Green	18,786	10.8%	7	8.3%
UKIP	20,182	4.6%	0	0%	Labour	33,873	13.5%	3	3.6%
Others [1]	24,302	5.5%	0	0%	Others [2]	17,764	7.1%	1	1.2%
Totals	438,226		9			251,351		84	
Turnout	66.8%					38.6%			

[1] Green, LCA, Independents, Others

[2] UKIP, LCA, Independents, Others

Settlements

Norfolk's county town and only city is Norwich, one of the largest settlements in England during the Norman era. Norwich is home to the University of East Anglia, and is the county's main business and culture centre. Other principal towns include the port-town of King's Lynn and the seaside resort and Broads gateway town of Great Yarmouth.

Based on the 2001 Census the county's largest centres of population are: Norwich (259,100) Great Yarmouth (71,700) Kings Lynn (43,100) Thetford (21,588) Dereham (15,659) [[Wymondham] (12,539) North Walsham (11,998) Cromer (7,749) Fakenham 7,357 Swaffham 6,935 There are also several smaller market towns: Aylsham, Downham Market,Diss,Holt, Hunstanton, and Sheringham.

Transport

Further information: Railways in Norfolk

Norfolk is one of the few counties in England that does not have a motorway. The A11 connects Norfolk to Cambridge and London via the M11. From the west there only two routes from Norfolk that have a direct link with the A1, the A47 which runs into the East Midlands and to Birmingham via Peterborough and the A17 which runs into the East Midlands via Lincolnshire. These two routes meet at King's Lynn which is also the starting place for the A10 which provides West Norfolk with a direct link to London via Ely, Cambridge and Hertford . The Great Eastern Main Line is a major railway from London Liverpool Street Station to Essex, Suffolk and Norfolk. Norwich International Airport, offers flights within Europe including a link to Amsterdam which offers onward flights throughout the world.

Dialect, accent and nickname

The Norfolk dialect is also known as "Broad Norfolk", although over the modern age much of the vocabulary and many of the phrases have died out due to a number of factors, such as radio, TV and people from other parts of the country coming to Norfolk. As a result, the speech of Norfolk is more of an accent than a dialect, though one part retained from the Norfolk dialect is the distinctive grammar of the region.

People from Norfolk are sometimes known as Norfolk Dumplings,[29] an allusion to the flour dumplings that were traditionally a significant part of the local diet.[30]

More cutting, perhaps, was the pejorative medical slang term "Normal for Norfolk",[31] alluding to the county's perceived status as an illiterate incestuous backwater. The term has never been official, and is now discredited, its use discouraged by the profession.

Tourism

Norfolk is a popular tourist destination and has several major examples of holiday attractions. There are many seaside resorts, including some of the finest British beaches, such as those at Great Yarmouth, Waxham, Cromer and Holkham bay. Norfolk is probably best known for the Broads and other areas of outstanding natural beauty and many areas of the coast are wild bird sanctuaries and reserves with some areas designated as National Parks such as the Norfolk Coast AONB. Tourists and locals enjoy the wide variety of monuments and historical buildings in both Norfolk and the city of Norwich.

The historic city of Norwich#Early English and Norman ConquestNorwich

The Norfolk Coast AONBNorfolk Coast at Cromer

The Norfolk Broads

The beach at Holkham Bay

The Queen's residence at Sandringham House in Sandringham, Norfolk provides an all year round tourist attraction whilst the coast and some rural areas are popular locations for people from the conurbations to purchase weekend holiday homes. Arthur Conan Doyle first conceived the idea for The Hound Of The Baskervilles whilst holidaying in Cromer with Bertram Fletcher Robinson after hearing local folklore tales regarding the mysterious hound known as Black Shuck.[32] [33]

Amusement parks and zoos

Norfolk has several amusement parks and zoos.

Thrigby Hall near Great Yarmouth was built in 1736 by Joshua Smith Esquire and features a zoo which houses a large Tiger enclosure, primate enclosures and the swamp house which has many Crocodiles and Alligators.

Pettitts Animal Adventure Park at Reedham is a park with a mix of Animals, rides and live entertainment shows.

Great Yarmouth Pleasure Beach is a free-entry theme park, hosting over 20 large rides as well as a crazy golf course, water attractions, children's rides and "white knuckle" rides.

BeWILDerwood is an award winning adventure park situated in the Norfolk Broads and is the setting for the book *A Boggle at BeWILDerwood* by local children's author Tom Blofeld.

Britannia Pier on the coast of Great Yarmouth has rides which include a ghost train. Also on the Pier is the famous Britannia Pier Theatre.

Banham Zoo is set amongst 35 acres (**unknown operator: u'strong'** m^2) of parkland and gardens with innovative enclosures providing sanctuary for almost 1000 animals including big cats, birds of prey, siamangs and shire horses. Its annual visitor attendance is in excess of 200,000 people, and has often been awarded the prize of Norfolk's Top Attraction, by numerous different organisations, including "Best Large Attraction" by *Tourism In Norfolk* in 2010.

Amazona Zoo is situated on 10 acres (**unknown operator: u'strong'** m^2) of derelict woodland and abandoned brick kilns on the outskirts of Cromer and is home to a range of tropical South American animals including jaguars, otters, monkeys and flamingos.

Amazonia is a tropical jungle environment in Great Yarmouth housing over 70 different species of reptiles including lizards, crocodiles, snakes, tortoises and terrapins.

Extreeme Adventure is a high ropes course built in some of the tallest trees in eastern England. in the 'New Wood' part of Weasenham Woods, Norfolk.

The Sea Life Centre in Great Yarmouth is One of the biggest sea life centres in the country. The Great Yarmouth centre is home to a tropical shark display, one resident of which is Britain's biggest shark 'Nobby' the Nurse Shark. The same display, with its walk-through underwater tunnel, also features the wreckage of a World War II aircraft. The centre also includes over 50 native species including shrimps, starfish, sharks, stingrays and Conger eels.

The Sea Life Sanctuary in Hunstanton is Norfolk's leading marine rescue centre and works both as a visitor attraction as well as a location for rescuing and rehabilitating sick and injured sea creatures found in the nearby Wash and North Sea. The attractions main features are similar to that of the Sea Life Centre in Great Yarmouth, albeit on a slightly smaller scale.

Theatres

The Pavilion Theatre(Cromer) is a 510-seater venue perched on the end of Cromer pier, best known for hosting the famous' end-of-the-pier' show, the Seaside Special. The theatre also presents a high-quality mix of comedy, music, dance, opera, musicals and community shows.

Britannia Pier Theatre (Great Yarmouth) host mainly popular comedy acts such as the Chuckle Brothers and Jim Davidson. the theatre seats 1200 seats and is one of the largest in Norfolk.

Theatre Royal (Norwich) has been on its present site for nearly 250 years, the Act of Parliament in the tenth year of the reign of George II having been rescinded in 1761. The 1300-seat theatre, the largest in the city, hosts a mix of national touring productions including musicals, dance, drama, family shows, stand-up comedians, opera and pop.

Norwich Playhouse (Norwich) is a superb venue in the heart of the city and one of the most modern performance spaces of its size in East Anglia. The theatre has a seating capacity of 300.

The Maddermarket Theatre (Norwich) opened in 1921 and was the first permanent recreation of an Elizabethan Theatre. The founder was Nugent Monck who had worked with William Poel. The Theatre is a world class Shakespearean style play house and has a seating capacity of 310.

Britannia Pier

Theatre Royal

Norwich Puppet Theatre (Norwich) was founded in 1979 by Ray and Joan DaSilva as a permanent base for their touring company and was first opened as a public venue in 1980, following the conversion of the medieval church of St. James in the heart of Norwich. Under subsequent artistic directors – Barry Smith and Luis Z. Boy – the theatre established its current pattern of operation. It is a nationally unique venue dedicated to puppetry, and currently houses a 185 seat raked auditorium, 50 seat Octagon Studio, workshops, an exhibition gallery, shop and licensed bar. It is the only theatre in the Eastern region with a year-round programme of family-centred entertainment.

Norwich Playhouse

The Garage studio theatre (Norwich) can seat up to 110 people in a range of different layouts. It can also be used for standing events and can accommodate up to 180 people. The high specification of equipment and design means that it is particularly versatile, and can be adapted to a variety of layouts offering a wide choice for performances or events.

The CCN Drama Centre [34] (Norwich) is in the grounds of City College Norwich (CCN), and has a massive stage. The theatre mainly plays host to Broadway style American musicals and has, in the past, presented shows such as Bugsy Malone and Copacabana and in May 2010 saw the highly successful, Thirst Productions Guys and Dolls. The theatre is raked and seats about 220 people.

The Sewell Barn Theatre (Norwich) is the smallest theatre in Norwich and has a seating capacity of just 100. The auditorium features raked seating on three sides of an open acting space. This unusual staging helps to draw the audience deeply into the performance.

The Norwich Arts Centre (Norwich) theatre opened in 1977 in St. Benedict's Street, and has a capacity of 290.

The Princess Theatre (Hunstanton) stands overlooking the Wash and green in the busy East Coast resort of Hunstanton. The Princess Theatre is a 472 seat venue dubbed as one of the friendliest theatres in the country by artists who have performed there. Open all year round, the theatre plays host to a wide variety of shows from comedy to drama, celebrity shows to music for all tastes and children's productions. The venue also has a six week summer season plus an annual Christmas pantomime.

Sheringham Little Theatre (Sheringham) provides intimate and comfortable seating for 180. The theatre programmes a wide variety of plays, musicals, music and also shows films.

The Gorleston Pavilion (Gorleston) is an original Edwardian building with a seating capacity of 300, situated on the Norfolk coast. The theatre stages plays, pantomimes, musicals, concerts as well as the popular 26 week Summer Season.

Notable people from Norfolk

Further information: Category:People from Norfolk

- Peter Bellamy, folk singer and musician, who was brought up in North Norfolk
- Henry Blofeld, Cricket commentator
- Henry Blogg, the UK's most decorated lifeboatman, who was from Cromer
- Francis Blomefield, Anglican rector, early topographical historian of Norfolk
- James Blunt, English acoustic folk rock singer-songwriter who was raised in Norfolk during his childhood
- James Blyth, author of weird fiction and crime mysteries, many of which are set in and around the Norfolk Broads
- Boudica, scourge of the occupying Roman Army in first century Britain and queen of the Iceni, British tribe occupying an area slightly larger than modern Norfolk

- Sir Thomas Browne, English Renaissance writer, physician and early archaeologist
- Martin Brundle, former motor-racing driver and now a commentator was born in King's Lynn
- Edward Bulwer-Lytton, 1st Baron Lytton, writer, born at Heydon
- Dave Bussey, former BBC Radio 2 and current BBC Radio Lincolnshire presenter
- Howard Carter, archaeologist who discovered Tutankhamun's tomb; his childhood was spent primarily in Swaffham
- Edith Cavell, a nurse executed by the Germans for aiding the escape of prisoners in World War I
- Nick Conrad, UK speech radio presenter, born in Norwich
- Deaf Havana, Alternative rockband from the King's Lynn area.
- Cathy Dennis, singer and songwriter, from Norwich
- Diana, Princess of Wales, first wife of Charles, Prince of Wales, was born and grew up near Sandringham
- Anthony Duckworth-Chad, landowner and Deputy Lord Lieutenant of Norfolk
- Sir James Dyson, inventor and entrepreneur, was born at Cromer, grew up at Holt and was educated at Gresham's School
- Bill (1916–1986), Brian (1922–2009), Eric (1914–1993), Geoff (1918–2004), John (1937–), and Justin (1961–) Edrich, cricketers
- E-Z Rollers, Drum and bass group was formed in Norwich.
- Nathan Fake, electronic dance music producer/DJ
- Pablo Fanque, equestrian and popular Victorian circus proprietor, whose 1843 poster advertisement inspired The Beatles song, Being for the Benefit of Mr. Kite!, born in Norwich
- Natasha and Ralph Firman, racing drivers, were both born and brought up in Norfolk and educated at Gresham's School
- Caroline Flack, television presenter, who grew up in East Wretham and went to school in Watton
- Margaret Fountaine, butterfly collector, was born in Norfolk, and her collection is housed in Norwich Castle Museum
- Elizabeth Fry, prominent 19th century Quaker prison reformer pictured on the Bank of England £5 note, born and raised in Norwich
- Stephen Fry, actor, comedian, writer, producer, director and author who was born in London and was brought up in the village of Booton near Reepham and also briefly attended Gresham's School. He now has a second home near King's Lynn.
- Samuel Fuller, signed the Mayflower Compact
- Claire Goose, actress who starred in *Casualty*, was raised in Norfolk
- Roderick Gordon, writer of Tunneles series.
- Ed Graham, drummer of Lowestoft band The Darkness, was born in Great Yarmouth
- Sienna Guillory, actress, from north Norfolk, who was educated at Gresham's School
- Sir Henry Rider Haggard, novelist, author of *She*, *King Solomon's Mines*, born Bradenham 1856 and lived after his marriage at Ditchingham
- Jake Humphrey, BBC presenter, spent most of his childhood in Norwich
- Andy Hunt, footballer, grew up in Ashill.
- Julian of Norwich, mediaeval mystic, born probably in Norwich in 1342; lived much of her life as a recluse in Norwich
- Robert Kett,leader of Kett's Rebellion in East Anglia 1549, from Wymondham
- Sid Kipper, Norfolk humourist, author, songwriter and singer
- Myleene Klass, former Hear'Say singer, comes from Gorleston
- Holly Lerski, singer and songwriter, former member of the band Angelou, grew up and resides in Norfolk
- Henry Leslie, actor and playwright, born 1830 at Walsoken.
- Samuel Lincoln, ancestor of U.S. President Abraham Lincoln

- Matthew Macfadyen, actor who starred in *Spooks*, was born in Great Yarmouth
- Ruth Madoc, actress, was born in Norwich
- Kenneth McKee, surgeon who pioneered hip replacement surgery techniques, lived in Tacolneston
- Roger Taylor, drummer of the rock band Queen was born in King's Lynn and spent the early part of his childhood in Norfolk.
- Danny Mills, footballer, born in Norwich.
- Horatio, Lord Nelson, Admiral and British hero who played a major role in the Battle of Trafalgar, born and schooled in Norfolk.
- Nimmo Twins, sketch comedy duo well known in Norfolk
- King Olav V of Norway, born at Flitcham on the Sandringham estate
- Beth Orton, singer/songwriter, was born in Dereham and raised in Norwich.
- Thomas Paine, philosopher, born in Thetford.
- Ronan Parke, Britain's Got Talent 2011 finalist and runner up.
- Margaret Paston, author of many of the Paston Letters, born 1423, lived at Gresham.
- Barry Pinches, snooker player who comes from Norwich.
- Matthew Pinsent, Olympic champion rower, was born in Holt.
- Prasutagus, 1st century king of the Iceni, who occupied roughly the area which is now Norfolk
- Philip Pullman, author, born in Norwich
- Anna Sewell, writer, author of *Black Beauty*, born at Great Yarmouth, lived part of her life at Old Catton near Norwich and buried at Lamas, near Buxton.
- Thomas Shadwell, playwright, satirist and Poet Laureate
- Allan Smethurst, 'The Singing Postman' who sang songs in his Norfolk dialect, was from Sheringham
- Hannah Spearritt, actress and former S Club 7 singer, who is from Gorleston
- Peter Trudgill, sociolinguist specialising in accents and dialects including his own native Norfolk dialect, was born and bred in Norwich
- George Vancouver, born King's Lynn. Captain and explorer in the Royal Navy
- Stella Vine, English artist, spent many of her early years in Norwich
- Sir Robert Walpole, first Earl of Orford, regarded as the first British prime minister
- Tim Westwood, rap DJ and Radio 1 presenter, grew up in and around Norwich
- Parson Woodforde, 18th century clergyman and diarist

People associated with Norfolk

The following people were not born or brought up in Norfolk but are long-term residents of Norfolk, are well known for living in Norfolk at some point in their lives, or have contributed in some significant way to the county.

- Verily Anderson, writer, lived in North Norfolk.
- Julian Assange, Australian publisher, journalist, writer, computer programmer, Internet activist and editor in chief of WikiLeaks, lived since 16 December 2010 in Ellingham Hall, the mansion of Vaughan Smith, under house arrest whilst fighting extradition to Sweden, before relocating to Kent in December 2011
- Bill Bryson, writer, has lived in the county since 2003.
- Adam Buxton, comedian and one half of Adam and Joe, moved to Norfolk in 2008
- Richard Condon (impresario), Theatre Royal, Norwich and Pavilion Theatre, Cromer Pier manager
- Revd Richard Enraght, 19th century clergyman, religious controversialist, Rector of St Swithun, Bintree
- Liza Goddard TV and stage actress, lives in the village of Syderstone.
- Trisha Goddard, TV personality, lives in Norwich and writes a column in the local newspaper the *Eastern Daily Press*.
- John Major British Prime Minister from 1990 to 1997, has a holiday home in Weybourne.
- Ronan Parke,child singer.

- Alan Partridge, fictional character associated with radio show *Norfolk Nights*
- Pocahontas, who lived at Heacham Hall for part of her life
- Martin Shaw, stage, television and film actor, is based in Norfolk.
- Delia Smith, cookery writer and major Norwich City Football Club shareholder
- John Wilson, angler, writer and broadcaster

See also

- Duke of Norfolk
- Earl of Norfolk
- List of Lord Lieutenants of Norfolk
- List of High Sheriffs of Norfolk
- Custos Rotulorum of Norfolk – List of Keepers of the Rolls
- Norfolk (UK Parliament constituency) – List of MPs for the Norfolk constituency
- List of places in Norfolk
- List of future transport developments in the East of England
- List of Parliamentary constituencies in Norfolk
- Norfolk Terrier
- Norwich Terrier
- Recreational walks in Norfolk
- Royal Norfolk Regiment

References

[1] Hierarchical list of the Nomenclature of Territorial Units for Statistics and the statistical regions of Europe (http://ec.europa.eu/comm/ eurostat/ramon/nuts/codelist_en.cfm?list=nuts) The European Commission, Statistical Office of the European Communities (retrieved 6 January 2008)

[2] Norfolk Government Statistics (http://66.102.9.104/search?q=cache:65bGq6uPxOIJ:www.norfolk.gov.uk/consumption/groups/public/ documents/general_resources/ncc041445.doc+norfolk+population+density&hl=en&ct=clnk&cd=1&gl=uk,)

[3] http://www.broads-authority.gov.uk/index.html

[4] John Barwell, n.d. " A History of Norfolk (http://www.norfolkbroads.com/guide/history.htm)."

[5] [www.geograph.org.uk/article/Medieval-Churches-in-Norfolk]

[6] "Voices of the Powerless: Boils and Buboes" (http://www.bbc.co.uk/radio4/history/voices/voices_salisbury.shtml). BBC Radio 4. 29 August 2002. . Retrieved 3 November 2008.

[7] Anon, 2002. Norfolk History (http://www.about-norfolk.com/about/county/norfolk history.htm).

[8] "Shoreline Management Plan" (http://web.archive.org/web/20080608020113/http://www.north-norfolk.gov.uk/coastal/default_5265. asp). north-norfolk.org. 22 February 2008. Archived from the original (http://www.north-norfolk.gov.uk/coastal/default_5265.asp) on 8 June 2008. . Retrieved 15 May 2008.

[9] Elliott, Valerie (29 March 2008). "Climate change: surrender a slab of Norfolk, say conservationists" (http://www.timesonline.co.uk/tol/ news/environment/article3642929.ece). *The Times* (London). . Retrieved 14 May 2008.

[10] Office for National Statistics, 2001. Regional Trends 26 (http://www.statistics.gov.uk/downloads/theme_compendia/ regional_trends_2001/rt36.pdf) ch:14.7 (PDF). Accessed 3 January 2006.

[11] Invest in Norfolk, Agriculture and Food (http://www.investinnorfolk.com/industry/display.jsp?dyn=industry_sector.20030625100834).

[12] http://www.newanglia.co.uk

[13] Hayes Computing Solutions (HCOMS) :: (http://www.hcoms.co.uk/free_wireless.php)

[14] University Campus Suffolk-Great Yarmouth college (http://www.ucs.ac.uk/SchoolsAndNetwork/Ourcampusnetwork/ UCSGreatYarmouth/UCS Great Yarmouth.aspx) Retrieved 30 June 2011

[15] "Local Government White Paper, Strong and Prosperous Communities" (http://www.norfolk.gov.uk/consumption/ idcplg?IdcService=SS_GET_PAGE&nodeId=3679). Norfolk County Council. . Retrieved 10 September 2009.

[16] "The business case for unitary Norwich" (http://www.norwich.gov.uk/site_files/pages/ City_Council__Unitary_Council__The_business_case.html). Norwich City Council. . Retrieved 13 February 2010.

[17] "Proposals for future unitary structures: Stakeholder consultation" (http://www.communities.gov.uk/index.asp?id=1509022). Communities and Local Government. . Retrieved 13 February 2010.

[18] "Our advice to the Secretary of State on unitary local government in Norfolk (PDF Document)" (http://www.lgbce.org.uk/__documents/ lgbce-documents/draftfinal-reports-and-consultation-papers/2009/norfolk-sr-final-dec09.pdf). The Boundary Committee. 7 December 2009.
.

[19] "Minister's Statement of 10 February 2010" (http://www.communities.gov.uk/news/localgovernment/1463780). Communities and Local Government. . Retrieved 13 February 2010.

[20] "Unitary Authorities" (http://www.publications.parliament.uk/pa/cm200809/cmhansrd/cm090224/halltext/90224h0001.htm). *House of Commons Hansard Debates*. Parliament of the United Kingdom. 24 February 2009. . Retrieved 13 February 2010.

[21] "Reaction to announcement on Local Government Reorganisation Announcement" (http://www.norfolk.gov.uk/consumption/ idcplg?IdcService=SS_GET_PAGE&ssDocName=NCC074317&ssSourceNodeId=&ssTargetNodeId=3018). *News Archive*. Norfolk County Council. 10 February 2010. . Retrieved 13 February 2010.

[22] "Peter Housden's letter in full" (http://www.edp24.co.uk/content/edp24/news/story.aspx?brand=EDPOnline&category=NewsSplash& tBrand=EDPOnline&tCategory=xDefault&itemid=NOED11 Feb 2010 20:18:28:560). *Eastern Daily Press*. 12 February 2010. .

[23] Shaun Lowthorpe (2 February 2010). "At last, a verdict on Norfolk councils' future" (http://www.edp24.co.uk/content/edp24/news/ story.aspx?brand=EDPOnline&category=NewsSplash&tBrand=EDPOnline&tCategory=xDefault&itemid=NOED02+Feb+2010+ 09:25:04:570). *Eastern Daily Pres.* .

[24] Lowthorpe, Shaun (14 May 2010). "Government chief moves to axe Norwich unitary plans" (http://www.edp24.co.uk/content/edp24/ news/story.aspx?brand=EDPOnline&category=News&tBrand=EDPOnline&tCategory=xDefault&itemid=NOED13 May 2010 19:12:43:450). *Eastern Daily Press.* .

[25] "Pickles stops unitary councils in Exeter, Norwich and Suffolk" (http://www.communities.gov.uk/news/corporate/159177711). Department for Communities and Local Government. . Retrieved 25 July 2010.

[26] "New bid to end unitary plans" (http://www.greatyarmouthmercury.co.uk/content/yarmouthmercury/news/story. aspx?brand=GYMOnline&category=news&tBrand=GYMonline&tCategory=news&itemid=NOED17 May 2010 09:50:04:040). *Yarmouth Mercury*. 17 May 2010. .

[27] "September by-elections for Exeter and Norwich" (http://www.bbc.co.uk/news/uk-england-10687867). BBC News. 19 July 2010. . Retrieved 19 July 2010.

[28] BBC News, Election 2009, Norfolk County Council. (http://news.bbc.co.uk/1/shared/bsp/hi/elections/local_council/09/html/3869. stm).

[29] Norfolk dumpling (http://www.norfolkdialect.com/lostintranslation/pages/sinkers.html) Retrieved 23 June 2011

[30] Nicknames (http://www.fromoldbooks.org/Grose-VulgarTongue/n/norfolk-dumpling.html) Retrieved 29 June 2011

[31] Doctor slang is a dying art (http://news.bbc.co.uk/1/hi/health/3159813.stm) *BBC News*, 18 August 2003

[32] "The Bookbag" (http://thebookbag.co.uk/reviews/index. php?title=Arthur_Conan_Doyle,_Sherlock_Holmes_and_Devon:_A_Complete_Tour_Guide_and_Companion_by_Brian_W_Pugh,_Paul_R_Spiring_and_Sadru Thebookbag.co.uk. Retrieved on 25 August 2011.

[33] "The District Messenger" (http://www.sherlock-holmes.org.uk/pdf/DM304.pdf). (PDF) . Retrieved on 25 August 2011.

[34] http://www.ccntheatre.webeden.co.uk

External links

- Norfolk (http://www.dmoz.org/Regional/Europe/United_Kingdom/England/Norfolk/) at the Open Directory Project
- Official Visitor site for the East of England (http://www.visiteastofengland.com)
- East of England Tourism (http://www.eet.org.uk)
- Eastern Daily Press – Norfolk Newspaper (http://www.edp24.co.uk)
- Norfolk County Council (http://www.norfolk.gov.uk)
- Norfolk tourism (official site) (http://www.visitnorfolk.co.uk)
- Norfolk Tourist Information & Articles (http://www.norfolkholidayguide.com)
- Norfolk Tourist Information (http://www.norfolktouristinformation.com)
- "60 Years of Change". A digital story, telling of the changes in a village school in rural Norfok (http://video. google.co.uk/videoplay?docid=1561406051370981651)
- Photos of Norfolk (http://www.scenicnorfolk.co.uk)
- Norfolk Rural Community Council, supports communities across Norfolk (http://www.norfolkrcc.org.uk)
- Norfolk E-Map Explorer – historical maps and aerial photographs of Norfolk (http://www.historic-maps. norfolk.gov.uk)
- Gallery of Norfolk (http://www.escapeandexplore.co.uk/galleries/norfolk.htm?view=link) – Photographs of Norfolk

- Photographs of North Norfolk (http://ninemalthousecourt.co.uk/photo_gallery.php)
- Norfolk Record Office (http://www.archives.norfolk.gov.uk/nroindex.htm) – Government agency that collects and preserves records of historical significance for the county of Norfolk and makes them accessible to the public. Useful for genealogical research.
- Norfolk County Council YouTube channel (http://uk.youtube.com/user/NorfolkCountyCouncil)

King's_Lynn

<table>
<tr><td colspan="3" align="center">King's Lynn</td></tr>
<tr><td colspan="3" align="center"></td></tr>
<tr><td colspan="3" align="center">From upper left: Custom House and South Gate, Vancouver Shopping District, Corn Exchange and King's Lynn railway station</td></tr>
<tr><td colspan="3" align="center">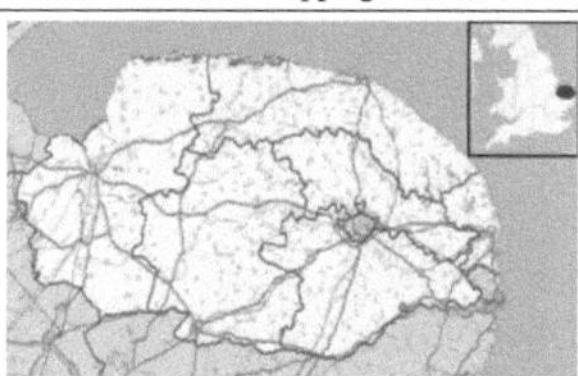</td></tr>
<tr><td colspan="3">King's Lynn

King's Lynn shown within Norfolk</td></tr>
<tr><td>Population</td><td colspan="2">Expression error: "42,800" must be numericTemplate:Infobox UK place/trap(2007)[1]</td></tr>
<tr><td>- London</td><td colspan="2">97 miles (unknown operator: u'strong' km)</td></tr>
<tr><td>District</td><td colspan="2">King's Lynn and West Norfolk</td></tr>
<tr><td>Shire county</td><td colspan="2">Norfolk</td></tr>
<tr><td>Region</td><td colspan="2">East</td></tr>
<tr><td>Country</td><td colspan="2">England</td></tr>
<tr><td>Sovereign state</td><td colspan="2">United Kingdom</td></tr>
<tr><td>Post town</td><td colspan="2">KING'S LYNN</td></tr>
<tr><td>Postcode district</td><td colspan="2">PE30</td></tr>
<tr><td>Dialling code</td><td colspan="2">01553</td></tr>
<tr><td>Police</td><td colspan="2">Norfolk</td></tr>
<tr><td>Fire</td><td colspan="2">Norfolk</td></tr>
<tr><td>Ambulance</td><td colspan="2">East of England</td></tr>
<tr><td>EU Parliament</td><td colspan="2">East of England</td></tr>
<tr><td>UK Parliament</td><td colspan="2">North West Norfolk</td></tr>
<tr><td>Website</td><td colspan="2">www.west-norfolk.gov.uk [2]</td></tr>
</table>

King's Lynn /ˌkɪŋzˈlɪn/ is a sea port and market town in the ceremonial county of Norfolk in the East of England. It is situated 97 miles (**unknown operator: u'strong'** km) north of London and 44 miles (**unknown operator:**

u'strong' km) west of Norwich.[3] The population of the town is 42,800.[1]

The town has two theatres, museums and other cultural and sporting venues. There are three secondary schools and one college. The service sector, information and communication technologies and creative industries, provide limited employment for the population of King's Lynn and the surrounding area.

History

Toponymy

The etymology of King's Lynn is uncertain. The name *Lynn* is said to be derived from the body of water near the town (the Great Ouse river as it prepares to enter the Wash): the Celtic word *Llyn*, means a lake; but the name is plausibly of Anglo-Saxon origin, from the word *Lean*, implying a tenure in fee or farm.[3]

For a time it was named *Len Episcopi* (Bishop's Lynn) while under the jurisdiction, both temporal and spiritual, of the Bishop of Norwich; but during the reign of Henry VIII of England it was surrendered to the crown, and it then assumed the name of *Lenne Regis*, or *King's Lynn*.[3]

In the Domesday Book, it is known as *Lun*, and *Lenn*; and is described as the property of the Bishop of Elmham, and the Archbishop of Canterbury.[3]

The town is and has been for generations generally known by its inhabitants and local people simply as *Lynn*. The city of Lynn, Massachusetts, just north of Boston, was named in 1637 in honour of its first official minister of religion, Samuel Whiting, who arrived at the new settlement from Lynn, Norfolk.[4]

Middle Ages

Lynn originated as a settlement on a constricted site to the south of the mouth of the River Great Ouse. Development began in the early 10th century, but was not recorded until the early 11th century.

In 1101, Bishop Herbert de Losinga of Thetford began the first mediaeval town between two rivers, the Purfleet to the north and Mill Fleet to the south, by commissioning St Margaret's Church and authorising a market.[5] In the same year, the Bishop granted the people of Lynn the right to hold a market on Saturday.[6]

Trade built up along the waterways that stretched inland from Lynn, and the town expanded between these two rivers.

Early modern

During the 14th century, King's Lynn ranked as the third most important port in England, behind Southampton and London. It was considered as important to England during the Middle Ages as Liverpool was during the Industrial Revolution. Sea trade with Europe was dominated by the Hanseatic League of ports; the transatlantic trade and the rise of England's western ports would not begin until the 17th century. The Trinity Guildhall was rebuilt in 1421 after a fire. It is debated whether the Guildhall of St George is the largest and oldest in England. In order to defend the town, walls and gatehouses were established, including the erection of the South Gate and East Gate.[7] The town retains two former warehouses of the Hanseatic League, in use between the 15th and 17th centuries. They are the only remaining buildings of the Hanseatic League in England.

Hanseatic Warehouse

In the first decade of the 16th century, Thoresby College was built by Thomas Thoresby to house priests of the Guild of The Holy Trinity in King's Lynn. The guild was incorporated in 1453 on the petition of its alderman, chaplain, four brethren and four sisters. The guildsmen were then licensed to found a chantry of chaplains to celebrate at the altar of Holy Trinity in Wisbech church and to grant to the chaplains lands in mortmain.[8] In 1524 King's Lynn was given a mayor and corporation. In 1537 the king took control of the town from the bishop. From then on it was called King's Lynn. However in the 16th century the town's two annual fairs were reduced to one. In 1534 a grammar school was founded. But in 1538 Henry VIII closed the Benedictine priory and the three friaries.

During the 16th century a piped water supply was created, although many could not afford to be connected: elm pipes carried water under the streets. Like all towns at that time King's Lynn suffered from outbreaks of plague: there were severe outbreaks in 1516, 1587, 1597, 1636 and 1665. But the 1665 outbreak proved to be the last. Fire was another hazard and in 1572 thatched roofs were banned to reduce the risk of fire. In 1642 came civil war between king and parliament. At first King's Lynn supported parliament, but in August 1643, after a change in government, the town changed sides. Parliament lost no time in sending an army to capture the town. King's Lynn was besieged for three weeks before it surrendered.

In 1683, the architect Henry Bell, who was once mayor of King's Lynn, built the Custom House. Bell also built the Duke's Head Inn, the North Runcton Church, and Stanhoe Hall. Bell's artistic inspiration was the result of travelling Europe as a young man.[9]

In the 16th and 17th centuries, the town's main export was grain. The town was no longer a major international port, although some iron and timber were still imported. Like other East Coast ports, King's Lynn suffered from the discovery of the Americas, which benefited ports on the West Coast of England. It was also affected by the growth of London which attracted the town's trade.

The Custom House

In the late 17th century, imports of wine from Spain, Portugal and France into King's Lynn boomed, and there was still an important coastal trade: at that time it was much cheaper to transport goods by water than by road, and thus many goods were shipped around the coast from one port to another. Large quantities of coal arrived in King's Lynn from North East England.

In the mid 17th century the fens were drained and turned into farmland. Vast amounts of farm produce were sent from King's Lynn to the growing market in London. King's Lynn was also still an important fishing port. Greenland Fishery House in Bridge Street was built in 1605. By the late 17th century shipbuilding had become an important industry in King's Lynn. A glass making industry also began in the late 17th century.

In the early 18th century Daniel Defoe said King's Lynn was: 'Beautiful, well built and well situated'. In the 18th century shipbuilding continued to thrive. So did associated industries such as sail making and rope making. Glass making continued to prosper. Brewing was another important industry. The first bank in King's Lynn opened in 1784.

On 28 September 1708, a seven year old boy, Michael Hammond and his 11 year old sister Ann Hammond were convicted of theft of a loaf of bread in King's Lynn. They were sentenced to death by hanging, a sentence which was carried out publicly near the South Gates of the town to make an example of them. At the time of the hangings, Sir Robert Walpole, generally regarded as the first Prime Minister of the United Kingdom, was Member of Parliament for King's Lynn.[10]

Modern

By the late 17th century, the town had begun to decline, and it was only rescued by the late arrival of railway services in 1847. These services were mainly provided by the Great Eastern Railway – subsequently London and North Eastern Railway. Trains ran from King's Lynn to Hunstanton, Dereham and Cambridge. The town was also served by the Midland and Great Northern Joint Railway, which had offices in the town at Austin Street, and an important station at South Lynn (now dismantled) which was also its operational control centre until this was relocated to Melton Constable. The M&GN lines were later closed to passengers in February 1959.

A photograph of the King's Lynn railway station

The town's amenities continued to improve into the 20th century. A museum opened in 1904, and a public library in 1905. The first cinema in King's Lynn, the Majestic Cinema, was officially opened on 23rdMay 1928; (this year is commemorated in the stained glass window at the front of the building) and the town council began regeneration of the town in the 1930s.

During World War I, Lynn was one of the first towns in Great Britain to be bombed from the air. On the night of January 19, 1915, the town was bombed by naval zeppelin L4 [11] commanded by Kapitain Lieutenant Magnus von Platen-Hallermund. Eleven bombs in all were dropped, both incendiary and high explosive, doing extensive damage, killing two people in Bentinck street and injuring several others.

When World War II began, it was assumed that King's Lynn would be safe from bombing, and many evacuees were sent there from London. However King's Lynn was not completely safe and suffered several air raids.

In 1962 King's Lynn became an overflow town for London, and the town's population increased. New estates were built at the Woottons and Gaywood. In the 1960s the town centre was redeveloped and many old buildings were destroyed. Lynnsport, a sports centre, opened in 1982. The Corn Exchange was converted to a theatre in 1996.

The brewing industry had died out by the 1950s but new industries came to King's Lynn: food canning in the 1930s and soup making in the 1950s. In the 1960s the council tried to attract new industries by building a new industrial estate at Hardwick. The new industries included light engineering, clothes and chemicals. Fishing remains an important industry.

In 1987, the town became the first in the UK to install town centre CCTV, though Bournemouth had previously used CCTV in non-central locations. The single crime most frequently prosecuted as a result of this comprehensive system is men urinating in public on their way home at night from pubs.

Contemporary

Since 2004, plans have been under way to regenerate the entire town. King's Lynn has undergone a multi-million pound regeneration scheme.

In 2005, the Vancouver Shopping Centre, (now since renamed the Vancouver quarter) originally built in the 1960s, was refurbished as part of the scheme, with a life expectancy of only 25 years according to the construction firm, and an extension is planned. A new award winning £6 million multi-storey car park was built.

King's Lynn, as viewed from across the River Great Ouse

To the south of town, a large area of brownfield land is being transformed into a housing estate locally known as Balamory after the colourful children's programme, and there were ambitions to build another housing estate alongside the River Nar but these were vehemently opposed by local opinion and the economic situation has seen this ambition stopped. There is also a business park, parkland, a school, shops and a new relief road in a £300 million+ scheme.

In 2006, King's Lynn became the United Kingdom's first member of The Hanse (Die Hanse), a network of towns and cities across Europe which historically belonged to the Hanseatic League. Originally this was a highly influential mediaeval trading association of merchant towns around the Baltic Sea and the North Sea, which contributed to the development of King's Lynn.[12]

The Borough Council commissioned a report by DTZ and accepted by the Borough Council published in 2008 which describes King's Lynn as a town with a workforce as being of "low value" and having a "low skills base". The town was further described as having a "poor lifestyle offer". The quality of services and amenities was described as "unattractive to higher value inward investors and professional employees with higher disposable incomes". Average earnings are well below regional and national levels, and a large number of jobs that do exist in tourism, leisure and hotels are both subject to seasonal fluctuations and are poorly paid. Education and workforce qualifications are described also as being below the national average. The borough ranks 150 out of 354 in terms of deprivation.[13]

In 2009, a proposal was submitted for the Campbell's Meadow factory site to be redeveloped to include a 5-hectare (**unknown operator: u'strong'**-acre) employment and business park, this plan had been rejected in favour of Sainsbury, but in June 2011 Tesco was given permission to build their store.[14] On 8 June 2010, Tesco unveiled its regeneration plans for the site that would cost £32 million, and might create 900 jobs overall.[15]

Tesco also pledged £4 million of improvements in other areas of the town. It planned to spend £1.6 million widening the Hardwick Road but the Sainsbury bid was preferred by the Council as it offered more benefits to the town.[15] Although now both stores will be constructed and as off August 2011 Tesco has started the redevelopment of the Campbell's site by slowly removing the current buildings which are full of asbestos for the construction of the new Tesco store which will be built behind the current Tesco Hardwick and will have twice the floor space it is due to open late 2012, as of the beginning of September no building work had been started on the new Sainsburys.

Sainsbury's has also had its £40 million plans for a new superstore opposite Tesco on the Pinguin Foods site, which is estimated to create 300 jobs and secure the future of Pinguin Foods in King's Lynn proposed and accepted by the town.[16] Pinguin Foods is releasing 12 acres (**unknown operator: u'strong'** m^2) of its 44-acre (**unknown operator: u'strong'** m^2) site, to accommodate the proposed store. Mortson Assets and Sainsbury's plan to create a new link road between *Scania Way* and *Queen Elizabeth Way* to improve access, allowing the industrial estate to expand and attract new employers. Sainsbury's will also keep their store open in the town centre. Sainsbury's has pledged £1.75 million for highways improvements and a further £7 million to invest in the Pinguin Foods factory.[15]

Destruction of iconic Campbell's tower

At 8am on the morning of Sunday January 15, 2012 the landmark, but by then derelict, Campbell's tower was demolished by competition winner Sarah Griffiths, whose father had died following a tragic accident at the factory 17 years previously. Mick Locke, 52, was fatally scalded by a blast of steam in 1995. An estimated three thousand people turned out to witness the towers final moments. It was Campbell's first UK factory when it opened in the 1950s, employing hundreds of local workers. At its peak in the early 1990s, it employed more than 700 workers. [17]

Campbell's tower in 2006

Governance

Historically part of the county of Norfolk, King's Lynn was made a county borough in 1883. The Borough of King's Lynn and West Norfolk was formed by the amalgamation of the Borough of King's Lynn, the Downham Market Urban District, the Hunstanton Urban District, the Docking Rural District, the Downham Rural District, the Freebridge Lynn Rural District, and the Marshland Rural District.[18]

Coat of arms

Coat of arms of King's Lynn and West Norfolk

The shield in the coat of arms of King's Lynn and West Norfolk is the arms of the ancient Borough of King's Lynn, which was recorded at the College of Arms in 1563. The shield shows the legend of Margaret of Antioch, who has been portrayed on the Seals of King's Lynn since the 13th century, and to whom the Parish Church is dedicated.[18]

The per chevron division and the addition of a bordure serve to make the shield distinct from its predecessor while retaining its medieval simplicity. The bordure also suggests the wider boundaries of the new authority, and the new shield is composed of seven parts to symbolise the seven authorities which were amalgamated.[18]

The gull depicted on the crest is a maritime reference. It appeared as a supporter in some representations of the arms, but officially it stands on a bollard in order to make it distinctive. It is supported with a crown or coronet like the King's Lynn supporter, and the lion in the crest of Downham Market Urban District Council coat of arms.

The crest of King's Lynn

The coronet refers to the Borough's royal connections. The cross held by the gull is an extension of the two in the shield, and the cross in the coat of arms of Freebridge Lynn Rural District.[18]

The supporters are based on the crest of the Hunstanton Urban District Council. The lion is a variation of the lions, or leopards, in the Royal Coat of Arms of the United Kingdom and its fish tail suggests the borough's links with the sea.[18]

The fish–lion is also the centre feature in the borough's badge, but here it is surrounded by a garland of oakleaves as a reference to the rural nature of much of the district. Oakleaves are also a feature of the coronet in the crest of the former Downham Market Urban District Council.[18]

.

Town twinning

King's Lynn has three twin towns:[19]

- Emmerich am Rhein, Germany[20]
- Jičín/Mladá Boleslav, Czech Republic
- Sandringham, Australia

Geography

Topography

King's Lynn is the northernmost settlement on the River Great Ouse, situated 97 miles (**unknown operator: u'strong'** km) north of London and 44 miles (**unknown operator: u'strong'** km) west of Norwich.[3] [21] [22] The town lies about 5 miles (**unknown operator: u'strong'** km) south of the Wash, an estuary on the northwest margin of East Anglia and 12 miles (**unknown operator: u'strong'** km) from the mouth of The Wash, an area subject to dangerous tides and shifting sandbanks. King's Lynn has an area of 11 square miles (**unknown operator: u'strong'** km^2).

The mouth of Gaywood River

The Great Ouse at Lynn is about 200 metres (**unknown operator: u'strong'** yd) wide and is the outfall for much of the drainage system of the Fens. The much smaller Gaywood River also flows through the town, joining the Great Ouse at the southern end of South Quay close to the town centre.

A small part, known as West Lynn, is on the west bank, and linked to the town centre by one of the oldest ferries in the country. Other districts of King's Lynn include the town centre, North Lynn, South Lynn, and Gaywood.

Climate

Along with the rest of the East of England, King's Lynn has a temperate climate. The annual mean daytime temperature is approximately 14 °C (**unknown operator: u'strong'** °F). January is the coldest month with mean minimum temperatures between 0 °C and 1 °C (32–33.8 °F). July and August are the warmest months, with mean daily maximum temperatures of approximately 21 °C (**unknown operator: u'strong'** °F).[23]

Two met office weather stations are in close proximity to Kings Lynn, Terrington St Clement, about 4 miles to the west of the town centre, and RAF Marham, about 10 miles due south south east.

The absolute maximum temperature at Terrington stands at 35.1c(95.2f)[24] recorded in August 2003, though in a more average year the warmest day will only reach 29.4c(84.9f),[25] with 13.8 days[26] in total attaining a temperature of 25.1c(77.2f) or more. Typically all these figures are marginally cooler than the southern half of the fens due to the not uncommon presence of an onshore sea breeze, and occasional haar/sea fog, particularly in early summer and late spring. However, with a strong enough offshore breeze, the area can be notably warm. Terrington (along with Cambridge Botanical Gardens) achieved the national highest temperature of 2007, 30.1c(86.2f)[27]

The absolute minimum at Terrington is -15.4c(4.3f),[28] set in January 1979. A total of 41.6 nights will report an air frost at Terrington and 51.9 nights at Marham.

Annual rainfall totals 621mm at Marham, and 599mm at Terrington,[29] with 1mm or more falling on 115 and 113 days,[30] respectively. All averages refer to the 30 year observation period 1971-2000.

Climate data for Marham

Month	Jan	Feb	Mar	Apr	May	Jun	Jul	Aug	Sep	Oct	Nov	Dec	Year
Average high °C (°F)	6.6	7.1	10.0	12.2	16.2	19.0	21.7	21.8	18.6	14.3	9.7	7.4	13.8
Average low °C (°F)	0.5	0.6	2.3	4.0	6.9	9.7	11.8	11.8	9.6	6.6	3.2	1.6	5.7
Precipitation mm (inches)	54.7 (2.154)	38.5 (1.516)	49.5 (1.949)	46.8 (1.843)	48.1 (1.894)	55.9 (2.201)	44.1 (1.736)	50.5 (1.988)	54.9 (2.161)	59.8 (2.354)	63.3 (2.492)	55.3 (2.177)	621.3 (24.461)
Mean monthly sunshine hours	53.6	73.2	101.7	150.6	204.3	191.1	202.7	192.8	139.8	109.7	69.0	48.1	1536.6
Source: Met Office[31]													

Parks

The town has several public parks, the largest one being The Walks, a historic 17 hectare urban park in the centre of King's Lynn. The Walks is the only surviving town walk in Norfolk from the 18th century. The Heritage Lottery Fund donated £4.3 million towards restoration on the park, including the addition of modern amenities. The Walks is also the location of The Red Mount, a Grade II-listed 15th century chapel. In 1998, the Walks was designated by English Heritage as a Grade II National historic park. The Walks as a whole had a different and earlier origin, in that it was at first conceived not as a municipal park, as one understands the term today, but as a single promenade for the citizens away from the smell, grime and bustle of the town centre. Harding's Pits is another public park and lies to the south of the town. It is an attractive informal area of open space with large public sculptures erected to reflect the history of the town. Harding's Pits is managed by local volunteers under a Management Company and has so far successfully fought off the Borough Council's attempts to turn it into an attenuation drain.[32]

Demography

As of 2007, King's Lynn has a population of 42,800.[1] According to Norfolk's 2007 census, King's Lynn, together with West Norfolk, has a population of 143,500, with an average population density of 1.00 persons per hectare.[1]

Economy

King's Lynn has always been a centre for the fishing and seafood industry (especially inshore prawns, shrimps and cockles). There have also been glass-making and small-scale engineering works (many fairground and steam engines were built here), and today, it is still the location for much agricultural-related industry including food processing. There are a number of chemical factories and the town retains a role as an import centre. It is a regional centre for what is still a sparsely populated part of England.

King's Lynn was the fastest growing port in Great Britain in 2008. The figures from the Department for Transport show that trade in the King's Lynn increased by 33 per cent.[33]

In 2008, the German Palm Group began to erect one of the world's largest paper machines. The machine was constructed by *Voith Paper*. With a web speed of up to 2000 m/min and a web width of 10.63m, it can produce 400.000 per year of newsprint paper. The production is based on 100% recycled paper. The start-up was on 21 August 2009.[34]

A panoramic view of the Vancouver Shopping District

The Port of King's Lynn has facilities for dry bulk cargo such as cereals and liquid bulk products such as petroleum products for Pace Petroleum. It also handles timber imported from Scandinavia and the Baltics, and has large handling sheds for steel imports.[35]

Retail

King's Lynn is the primary retail centre in West Norfolk, as well as being the principal centre for people living outside the border of West Norfolk. The town centre is dominated by budget shops reflecting the spending power of much of the population. The town centre fulfils a leisure role with entertainment centres, bars and restaurants, and has a range of service functions. There are around 5,300 retailing jobs.[36]

The Vancouver Shopping District at night.

The town centre has 73,000 square metres of retail floorspace in 347 shops, which is greater than the comparable centres of Bury St Edmunds and Boston. However, whilst the percentage of floorspace in comparison shopping and that occupied by multiple retailers is above the national average, King's Lynn offers limited range of choice.[36]

Tourism

Tourism in King's Lynn is a minor industry and attracts a relatively tiny amount of tourists each year.

Transport

Major routes

South Transport Project

A £7 million program to redevelop King's Lynn's Town Centre's infrastructure is due for completion in 2011. The majority of the money is provided by the Community Infrastructure Fund. The department program is a collection of smaller developments which are detailed below.[37]

A cycle and bus route between the town centre and South Lynn started in June 2010 at a cost of £850,000. The route will be 720 metres long, running from Morston Drift to Millfleet, with buses travelling in both directions along it. It will also feature a separate path for pedestrians and bicycles, this path will meet the bus route when crossing the Nar sluice. As part of this development, the Millfleet - St James' Road junction will be developed to better accommodate the envisioned increased bus and bike traffic.[37]

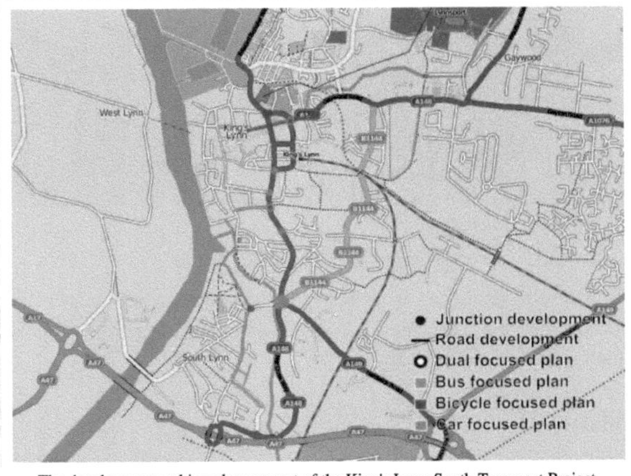

The developments taking place as part of the King's Lynn South Transport Project.

A contraflow lane for bicycles was proposed but will not be built along Norfolk Street from Albert Street to Blackfriars Road, this would have included a development of the Norfolk Road - Railway Road junction to better accommodate buses and bicycles. Similar work would have taken place at the Norfolk Street - Littleport Street junction so that buses do not get caught in the town centre gyratory system.[37]

Bus priority measures will be added to four sets of traffic lights along St James' Road. This is being undertaken to give buses quicker access to the town centre and normalise journey times.[37]

Southgates Roundabout has also been redeveloped. Many of the approach roads will be widened in the run up to the junction and the road markings will be redone in an attempt to improve lane discipline. Southgates Roundabout is a noted congestion hotspot by the county council and thus targeted by this scheme as a point to be developed.[37]

Other small developments are taking place to make junctions more car-friendly.[37]

Buses

Norfolk Green provides a regular bus services to many surrounding towns and villages around Norfolk. The town is connected to the local cities of Norwich and Peterborough via the A47 and to Cambridge via the A10.

Rail

King's Lynn railway station is the only rail line providing rail transportation to King's Lynn, and is the terminus of the Fen Line. The station provides connections to Ely, Cambridge and London King's Cross. It is the only remaining station of several the town once hosted. South Lynn railway station closed to passengers in 1959, and the railway line to the Hunstanton railway station was closed in 1969.

West Norfolk Council are considering reopening the railway route between the King's Lynn railway station and the Hunstanton railway station. The possibility of reinstating the line was proposed at a meeting of the council's Regeneration and

A Class 365 train at King's Lynn railway station in November 2009.

Environment Panel on 29 October 2008. The re-opening of the route was last discussed in the 1990s. The environmental case for reviving the line and relieving road congestion in and around Hunstanton is considered to be even stronger.[38]

Media

King's Lynn has one main local newspaper, the Lynn News. It is a twice weekly newspaper, largely based on advertising, owned by East Midlands Newspapers. The Lynn News has two local sister newspapers; the Peterborough Evening Telegraph and the Fenland Citizen.[39]

The local college runs a web-based TV station from the media department's students, entitled SpringboardTV.com and runs a little awards ceremony at the end of every academic year. This year, it won an award for most outstanding media department within the entire country of the United Kingdom.

For television, King's Lynn is served by BBC Yorkshire and Lincolnshire and ITV Yorkshire, and also served by BBC East and ITV Anglia.

King's Lynn has one locally broadcast radio station, KL.FM 96.7, a commercial radio station with local programmes.[40]

Education

King's Lynn has three secondary schools; King Edward VII School, the King's Lynn Academy, and Springwood High School. The former is known, academically, for its physical education department.[41] [41] [42] [43]

The town contains one further education college, the College of West Anglia. It was founded in 1894, as the King's Lynn Technical School. In 1973, it was renamed The Norfolk College of Arts and Technology, and in 1998, it merged with the Cambridgeshire College of Agriculture and Horticulture, which added two campuses in Wisbech and Milton. In April 2006, the College merged with the Isle College in Wisbech to form the College of West Anglia.[44]

Culture

Arts

Lady Ruth Fermoy, an accomplished concert pianist, moved to King's Lynn in 1931, as the bride of Lord Edmund Fermoy, who was to become the mayor and MP of the town. She demonstrated her affection for the town by organising concerts to give the local people the chance to listen to professional music of the highest standard.[45]

In 1951 to complement the Festival of Britain, Lady Fermoy organised the King's Lynn Festival of the Arts. She was a close friend and lady-in-waiting to the Queen – later to become Queen Elizabeth The Queen Mother – who agreed to become the festival's patron, and in July 1951 officially opened the restored St George's Guildhall. The Queen Mother was an enthusiastic and active supporter who remained the festival's patron until her death in March 2002.[45]

King's Lynn Literature Festivals

The King's Lynn Literature Festivals are held during a single weekend in March (Fiction) and September (Poetry) each year, usually in the town hall.[46]

Museum

There is a small museum of the former life of the North End fishermen at True's Yard. It includes cottages and a former smokehouse.

Entertainment

Festival Too is held on the Tuesday Market Place every summer. Past performers include Midge Ure, Deacon Blue, Suzi Quatro, 10cc, Mungo Jerry, The Human League, Buzzcocks, M People, Atomic Kitten, Kieran Woodcock, S Club, and Beverley Knight.

Local bands

- Deaf Havana
- Rebelation
- Vanilla Pod
- Faintest Idea
- Captain Black No Stars
- AKAKen
- Ringo Pacino
- Officer Gotcha

The Majestic Cinema, located in the town centre, is the town's only cinema.

King's Lynn's main venue for concerts, stand-up comedy shows and other live events is the Corn Exchange, located on Tuesday Market Place. With many smaller venues such as Bar red and The Wenns supporting the vibrant local music scene as well as many unsigned acts from other parts of the country. [47]

Mart

During the 16th century, King's Lynn's Tuesday Market Place hosted two important trade fairs which attracted visitors from as far as Italy and Germany. As the importance of trade fairs declined, the Mart's nature changed to become a funfair, and was reduced to a single annual event that takes place on 14 February (Valentine's Day), and lasts an average of 14 days.

The Mart is also a memorial to the work of Frederick Savage, a man who worked in partnership with the Showmen's Guild of Great Britain to develop new attractions.[48]

The Mart on the Tuesday Market Place.

Sport

King's Lynn F.C. club (nicknamed "The Linnets") played football in the Northern Premier League. It had its ground at The Walks Stadium on Tennyson Road. It was officially wound up in the High Court in December 2009. In 2010 it re-formed with the new name King's Lynn Town F.C..

King's Lynn also has a motorcycle speedway team, the King's Lynn Stars, who race at the Norfolk Arena on Saddlebow Road. The track has operated since 1965 when it operated on an open licence. Speedway type events were staged at the stadium in the 1950s.

The basketball team Lynn Nets, is also based in King's Lynn.

Wales and Scarlets rugby union player George North was also born here.

Notable people

- Captain George Vancouver (born 1757)
- Margery Kempe (born c. 1373), pilgrim and mystic, produced arguably the first autobiography in the English language.
- Frances Burney (born 1752), writer, produced the classic work *Evelina*.
- Walter Dexter RBA 1876–1958, artist, Lived in the Valiant sailor on Nelson street until he was tragically killed walking across the Saturday market place. His most well known work was The Carpenters Workshop, painted in 1904.[49]
- Florence Green (born 1901), one of Britain's oldest people until her death in 2012 and the last surviving British World War I veteran. She was born in London but lived in King's Lynn from 1920.[50]
- Roger Taylor, drummer with the rock band Queen, was born at King's Lynn in 1949, although, he moved to Cornwall as a child during the 1950s.[51] [52]
- Martin Brundle, former racing driver, was born in the town in 1959, as was his son Alex (also a racing driver) in 1990.[53] [54]
- Claire Goose (born 1975), actress, grew up in the town having been born in Edinburgh.[55]
- Nick Aldis (born 1986), Professional Wrestler with TNA, known as Magnus.
- Arthur Lowe, star of Dad's Army, lived here.
- George North (born 1992), Wales rugby union international, was born in King's Lynn, but moved with his family to the Welsh Isle of Anglesey when he was two years old.
- Stephen Fry Actor, comedian, writer, and presenter as well as director of Norwich football club famously lives near King's Lynn.
- Deaf Havana James Veck-Gilodi, Lee Wilson, Tom Ogden, Chris Pennells. English Post Hardcore Rock band formed in King's Lynn.

- Martin Saggers England cricketer grew up here.
- Jody Cundy Paralympic cyclist grew up here.

See also

- List of buildings in King's Lynn
- List of people from King's Lynn
- King's Lynn Docks
- King's Lynn Power Station

References

[1] "CAn Overview of King's Lynn and West Norfolk - Part 1" (http://www.norfolk.gov.uk/consumption/groups/public/documents/ general_resources/ncc041459.pdf) (PDF). 2007. p. 2. . Retrieved 2010-05-15.

[2] http://www.west-norfolk.gov.uk/

[3] Lewis, Samuel (1848). "Lynn, or Lynn-Regis". *A Topographical Dictionary of England*. pp. 203–208.

[4] http://www.ci.lynn.ma.us/aboutlynn_history.shtml

[5] "History and Heritage of King's Lynn" (http://www.west-norfolk.gov.uk/default.aspx?page=21900). Borough Council of King's Lynn & West Norfolk. . Retrieved 2010-05-25.

[6] Lambert, Tim. "A history of King's Lynn" (http://www.localhistories.org/Kingslynn.html). . Retrieved 2010-06-02.

[7] "King's Lynn" (http://www.poppyland.co.uk/index.php?s=KINGS_LYNN). *Poppyland Publishing*. . Retrieved 2010-06-08.

[8] Pugh, R.B. (2002). "Guild of the Holy Trinity". *A History of the County of Cambridge and the Isle of Ely: Volume 4: City of Ely; Ely, N. and S. Witchford and Wisbech Hundreds*. pp. 255–256.

[9] "Custom House, King's Lynn" (http://www.edp24.co.uk/content/edp24/norfolk-life/norfolk-history/content/22CustomHouse.aspx). *Eastern Daily Press*. . Retrieved 2010-06-08.

[10] "Death penalty abolished" (http://www.information-britain.co.uk/famdates.php?id=282). 2007-04-01. . Retrieved 2010-06-02.

[11] http://blog.ezep.de/2011/02/zeppelin-l4-lz-27-crashed-at-denmark-fuel-shortage/

[12] "King's Lynn, a Hanse League Member" (http://www.west-norfolk.gov.uk/default.aspx?page=23286). King's Lynn and West Norfolk Borough Council Website. . Retrieved 2007-01-15.

[13] "Economic Impact Assessment of King's Lynn Marina" (http://www.west-norfolk.gov.uk/pdf/The Economic Impact Assessment.pdf) (PDF). 2007-06. . Retrieved 2010-08-22.

[14] "Welcome to Campbells Meadow" (http://www.campbellsmeadow.co.uk/01_01_investment.html). Tesco. 2009. . Retrieved 2010-05-25.

[15] "Supermarket giants battle it out for Hardwick contract" (http://www.lynnnews.co.uk/news/features/ supermarket_giants_battle_it_out_for_hardwick_contract_1_648674). Lynn News. 2010-06-08. . Retrieved 2010-06-08.

[16] "A New Sainsbury's for King's Lynn" (http://www.sainsburys-kingslynn.co.uk/). Sainsbury's. 2009. . Retrieved 2010-06-11.

[17] http://www.edp24.co.uk/news/video_thousands_gather_to_watch_campbell_s_tower_demolished_in_king_s_lynn_1_1177765

[18] "King's Lynn and West Norfolk County Council" (http://www.civicheraldry.co.uk/east_anglia_essex.html#kings lynn and west norfolk bc). Civic Heraldry. . Retrieved 2010-05-27.

[19] http://www.hanse.org/en/hanseatic-cities/kings_lynn-tourism

[20] "Stadt Emmerich am Rhein" (http://www.emmerich.de/). *Emmerich am Rhein*. . Retrieved 2010-05-15.

[21] *OS Explorer Map 250 - Norfolk Coast West*. Ordnance Survey. 2002. ISBN 0-319-21886-4.

[22] *OS Explorer Map 236 - King's Lynn, Downham Market & Swaffham*. Ordnance Survey. 1999. ISBN 0-319-21867-8.

[23] "Met Office: Climate averages 1971–2000" (http://www.metoffice.gov.uk/climate/uk/averages/19712000/sites/marham.html). *Met Office*. 2000. . Retrieved 2010-05-15.

[24] "2003 maximum" (http://eca.knmi.nl/utils/monitordetail.php?seasonid=14&year=2003&indexid=TXx&stationid=1840). . Retrieved 2011-02-28.

[25] "1971-00 average warmest day" (http://eca.knmi.nl/utils/calcdetail.php?seasonid=0&periodid=1971-2000&indexid=TXx& stationid=1840). . Retrieved 2011-02-28.

[26] "1971-00 >25c days" (http://eca.knmi.nl/utils/calcdetail.php?seasonid=0&periodid=1971-2000&indexid=SU&stationid=1840). . Retrieved 2011-02-28.

[27] "2007 maximum" (http://www.metoffice.gov.uk/climate/uk/2007/august.html). . Retrieved 2011-02-28.

[28] "1979 minimum" (http://eca.knmi.nl/utils/monitordetail.php?seasonid=7&year=1979&indexid=TNn&stationid=1840). . Retrieved 2011-02-28.

[29] "average rainfall" (http://eca.knmi.nl/utils/calcdetail.php?seasonid=0&periodid=1971-2000&indexid=RR&stationid=1840). . Retrieved 2011-02-28.

[30] "average raindays" (http://eca.knmi.nl/utils/calcdetail.php?seasonid=0&periodid=1971-2000&indexid=RR1&stationid=1840). . Retrieved 2011-02-28.

[31] "Marham 1971–2000 climate averages" (http://www.metoffice.gov.uk/climate/uk/averages/19712000/sites/marham.html). Met Office. . Retrieved July 1, 2010.

[32] "The Walks" (http://www.west-norfolk.gov.uk/default.aspx?page=22720). *King's Lynn Online*. . Retrieved 2010-06-06.

[33] "King's Lynn is fastest growing port in Britain" (http://www.businessweekly.co.uk/2009110435788/travel-and-transport/ king-s-lynn-is-fastest-growing-port-in-britain.html). Business Weekly. 2009-11-04. . Retrieved 2010-06-17.

[34] "Standorte - King's Lynn - English" (http://www.papierfabrik-palm.de/ISY/index.php?call=en_kingslynn&sprache=en). The Palm Group. . Retrieved 2010-05-25.

[35] "Port of Kings Lynn: Comodities" (http://www.abports.co.uk/custinfo/ports/kings/commodities.htm). Associated British Ports. . Retrieved 2010-05-25.

[36] "The Vision for King's Lynn 2000-2023" (http://www.west-norfolk.gov.uk/pdf/The Vision statement.pdf) (PDF). 2004-04. p. 2. . Retrieved 2010-06-02.

[37] "King's Lynn South Transport Major Scheme" (http://www.norfolk.gov.uk/consumption/groups/public/documents/article/ncc063857. pdf). . Retrieved 2009-11-17.

[38] "Lynn-Hunstanton rail line re-opening hope revived" (http://www.lynnnews.co.uk/news/ lynn_hunstanton_rail_line_re_opening_hope_revived_1_521513). Lynn News. 2008-10-29. . Retrieved 2010-06-08.

[39] "Lynn News" (http://www.lynnnews.co.uk/). *Lynn News*. . Retrieved 2010-05-27.

[40] "West Norfolk's KL.FM 96.7" (http://www.klfm967.co.uk/). *KL.FM 96.7*. . Retrieved 2010-05-25.

[41] "King Edward VII School" (http://www.kingedward.norfolk.sch.uk/home.php). *King Edward VII School*. . Retrieved 2010-05-15.

[42] "The Park High School" (http://www.park.norfolk.sch.uk/). *The Park High School*. . Retrieved 2010-05-15.

[43] "Springwood High School" (http://www.springwood.norfolk.sch.uk/). *Springwood High School*. . Retrieved 2010-05-15.

[44] "History of College" (http://www.col-westanglia.ac.uk/about-history.html). *College of West Anglia*. . Retrieved 2010-05-25.

[45] "History of the King's Lynn Festival" (http://www.kingslynnfestival.org.uk/pages/history-amp-hall-of-fame.php). *kingslynnfestival.org.uk*. . Retrieved 2010-06-06.

[46] http://www.lynnlitfests.com/Welcome.html

[47] http://www.kingslynncornexchange.co.uk/

[48] "King's Lynn History" (http://www.kingslynnonline.com/kings-lynn-history.htm). *Borough Council of King's Lynn and West Norfolk*. . Retrieved 2010-06-06.

[49] http://www.google.co.uk/search?source=ig&hl=en&rlz=1G1GGLQ_ENGB457&q=Walter+Dexter&btnG=Google+Search

[50] Britten, Nick (16 January 2010). "108-year-old woman emerges as Britain's oldest first World War veteran" (http://www.telegraph.co.uk/ news/uknews/6996800/108-year-old-woman-emerges-as-Britains-oldest-first-World-War-veteran.html). *The Daily Telegraph* (London). .

[51] (http://www.information-britain.co.uk/famousbrits.php?id=1679)

[52] (http://www.wewillrockyou.co.uk/show/cast-and-creatives/roger-taylor/)

[53] (http://en.espnf1.com/f1/motorsport/driver/1093.html)

[54] (http://www.castroldriverrankings.com/profile/alex-brundle/1990080700)

[55] (http://www.holby.tv/db/index.php?id=38,93,0,0,1,0)

External links

- King's Lynn travel guide from Wikitravel
- King's Lynn (http://www.dmoz.org//Regional/Europe/United_Kingdom/England/Norfolk/King's_Lynn//) at the Open Directory Project
- Information from Genuki Norfolk (http://www.origins.org.uk/genuki/NFK/places/k/kings_lynn/)
- History of medieval Lynn (http://www.trytel.com/~tristan/towns/lynn1.html)
- Borough Council of King's Lynn and West Norfolk (http://www.west-norfolk.gov.uk)

Great_Yarmouth

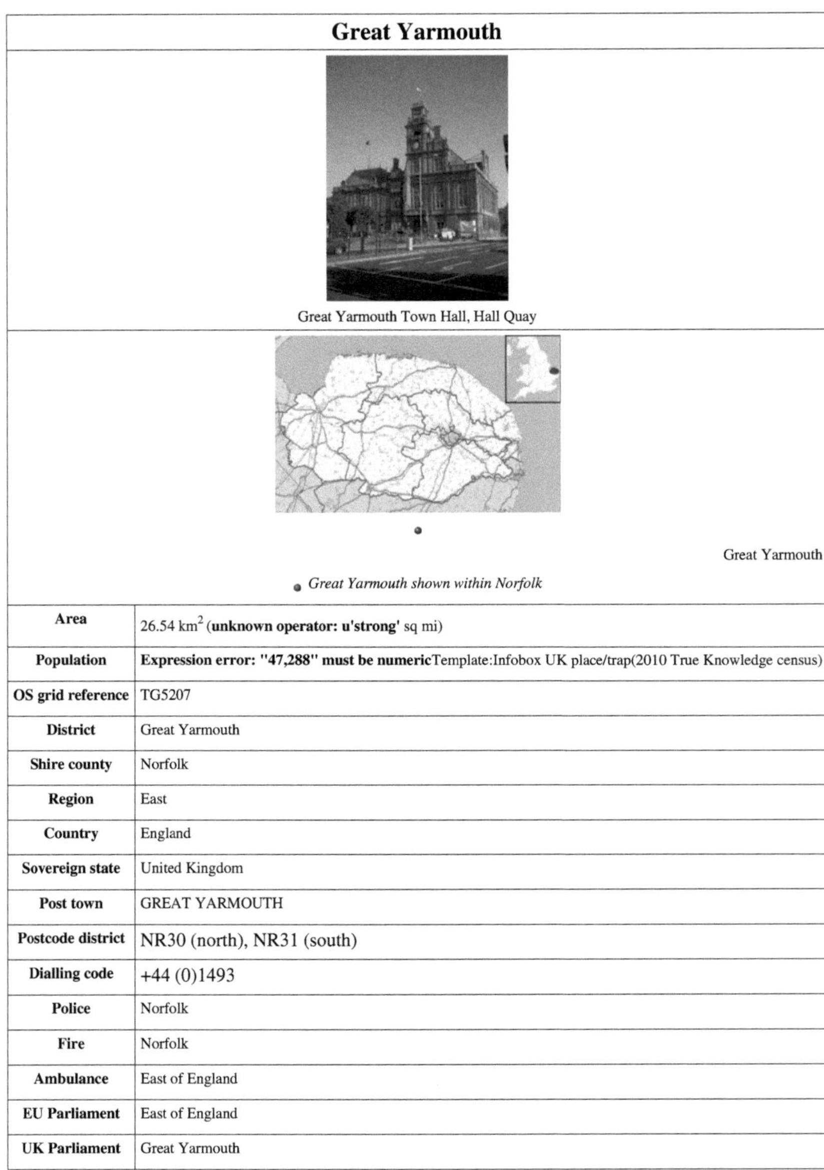

<table>
<tr><td colspan="2" align="center">Great Yarmouth</td></tr>
<tr><td colspan="2" align="center">Great Yarmouth Town Hall, Hall Quay</td></tr>
<tr><td colspan="2" align="center">Great Yarmouth
Great Yarmouth shown within Norfolk</td></tr>
<tr><td>Area</td><td>26.54 km² (unknown operator: u'strong' sq mi)</td></tr>
<tr><td>Population</td><td>Expression error: "47,288" must be numericTemplate:Infobox UK place/trap(2010 True Knowledge census)</td></tr>
<tr><td>OS grid reference</td><td>TG5207</td></tr>
<tr><td>District</td><td>Great Yarmouth</td></tr>
<tr><td>Shire county</td><td>Norfolk</td></tr>
<tr><td>Region</td><td>East</td></tr>
<tr><td>Country</td><td>England</td></tr>
<tr><td>Sovereign state</td><td>United Kingdom</td></tr>
<tr><td>Post town</td><td>GREAT YARMOUTH</td></tr>
<tr><td>Postcode district</td><td>NR30 (north), NR31 (south)</td></tr>
<tr><td>Dialling code</td><td>+44 (0)1493</td></tr>
<tr><td>Police</td><td>Norfolk</td></tr>
<tr><td>Fire</td><td>Norfolk</td></tr>
<tr><td>Ambulance</td><td>East of England</td></tr>
<tr><td>EU Parliament</td><td>East of England</td></tr>
<tr><td>UK Parliament</td><td>Great Yarmouth</td></tr>
</table>

Great Yarmouth, often known to locals as **Yarmouth**, is a coastal town in Norfolk, England. It is located at the mouth of the River Yare, 20 miles (**unknown operator: u'strong'** km) east of Norwich.[1]

It has been a seaside resort since 1760, and is the gateway from the Norfolk Broads to the sea. For hundreds of years it was a major fishing port, depending mainly on herring fishery, but its fishing industry suffered a steep decline and has now all but disappeared.[2] The discovery of oil in the North Sea in the 1960s led to a flourishing oil rig supply industry, and today it services offshore natural gas rigs. More recently, the development of renewable energy sources, especially offshore wind power, has created further opportunities for support services. A wind farm of 30 generators is within sight of the town on the Scroby Sands.

The town has a beach and two piers.

Geography and demography

The town itself is on a thin spit sandwiched between the North Sea and River Yare. Its well known features include the historic rows (narrow streets) and the main tourist sector on the seafront. The area is linked to Gorleston, Cobholm and Southtown by Haven Bridge and to the A47, A149 and A12 by the Breydon Bridge.

The unparished urban area that makes up the town of Great Yarmouth has an area of 21.5 km^2 (**unknown operator: u'strong'** sq mi) and according to the Office for National Statistics in 2002 had a population of 47,288. It is the main town in the larger Borough of Great Yarmouth.[3] The ONS identify a Great Yarmouth Urban Area, which has a population of 66,788, including the sub-areas of Caister-on-Sea (8,756) and Great Yarmouth (58,032). The wider borough of Great Yarmouth has a population of around 92,500.

History

Great Yarmouth (Gernemwa, Yernemuth) lies near the site of the Roman fort camp of Gariannonum at the mouth of the River Yare. Its situation having attracted fishermen from the Cinque Ports, a permanent settlement was made, and the town numbered 70 burgesses before the Norman Conquest. Henry I placed it under the rule of a reeve.

The charter of King John (1208), which gave his burgesses of Yarmouth general liberties according to the customs of Oxford, a gild merchant and weekly hustings, was amplified by several later charters asserting the rights of the borough against Little Yarmouth and Gorleston. A thirteenth century charter was granted by Henry III (1207–1272) to the town of Great Yarmouth. The town is bound to send to the sheriffs of Norwich every year *one hundred herrings, baked in twenty four pasties*, which the sheriffs are to deliver to the lord of the manor of East Carlton who is then to convey them to the King.[4]

In 1552 Elizabeth granted a charter of admiralty jurisdiction, later confirmed and extended by James I. In 1668 Charles II incorporated Little Yarmouth in the borough by a charter which with one brief exception remained in force until 1703, when Anne replaced the two bailiffs by a mayor. In the early 18th Century Yarmouth, as a thriving herring port, was vividly and admiringly described several times in Daniel Defoe's travel journals, in part as follows:[5]

The 133 foot-tall Britannia Monument, built in 1817.

> Yarmouth is an antient town, much older than Norwich; and at present, tho' not standing on so much ground, yet better built; much more compleat; for number of inhabitants, not much inferior; and for wealth, trade, and advantage of its situation, infinitely superior to Norwich.

> It is plac'd on a peninsula between the River Yare and the sea; the two last lying parallel to one another, and the town in the middle: The river lies on the west-side of the town, and being grown very large and deep, by a

conflux of all the rivers on this side the county, forms the haven; and the town facing to the west also, and open to the river, makes the finest key in England, if not in Europe, not inferior even to that of Marseilles itself.

The ships ride here so close, and as it were, keeping up one another, with their head-fasts on shore, that for half a mile together, they go cross the stream with their bolsprits over the land, their bowes, or heads, touching the very wharf; so that one may walk from ship to ship as on a floating bridge, all along by the shore-side: The key reaching from the drawbridge almost to the south-gate, is so spacious and wide, that in some places 'tis near one hundred yards from the houses to the wharf. In this pleasant and agreeable range of houses are some very magnificent buildings, and among the rest, the custom-house and town-hall, and some merchants houses, which look like little palaces, rather than the dwelling-houses of private men.

The greatest defect of this beautiful town, seems to be, that tho' it is very rich and encreasing in wealth and trade, and consequently in people, there is not room to enlarge the town by building; which would be certainly done much more than it is, but that the river on the land-side prescribes them, except at the north end without the gate....

A grammar school was founded in 1551, when the great hall of the old hospital, founded in the reign of Edward I by Thomas Fastolfe, was appropriated to its use. It was closed from 1757 to 1860, was re-established by the charity trustees, and settled in new buildings in 1872.

From 1808 to 1814 the Admiralty in London could communicate with its ships in the port of Great Yarmouth by a shutter telegraph chain.

The town was the site of a bridge disaster and drowning tragedy on 2 May 1845 when a suspension bridge crowded with children collapsed under the weight killing 79. They had gathered to watch a clown in a barrel being pulled by geese down the river. As he passed under the bridge the weight shifted, causing the chains on the south side to snap, tipping over the bridge deck.[6]

Great Yarmouth had an electric tramway system from 1902 to 1933.

During World War I Great Yarmouth suffered the first aerial bombardment in the UK, by Zeppelin *L3* on 19 January 1915. That same year on 15 August, Ernest Martin Jehan became the first and only man to sink a steel submarine with a sail rigged Q-ship, this off the coast of Great Yarmouth. It was also bombarded by the German Navy on 24 April 1916.

The town suffered Luftwaffe bombing during World War II as it was the last significant place German bombers could drop bombs before returning home, but much is left of the old town including the original 2000m protective mediaeval wall, of which two-thirds has survived. Of the 18 towers, 11 are left. On the South Quay is a 17th century Merchant's House, as well as Tudor, Georgian and Victorian buildings. Behind South Quay is a maze of alleys and lanes known as "The Rows". Originally there were 145. Despite war damage, several have remained.

Britannia Pier in 1930.

The northern section of the two-mile (3 km) A47 Great Yarmouth Western Bypass opened in March 1986, and the southern section in May 1985. It is now the A12.

More recently flooding has been a problem, the town flooding four times in 2006. In September 2006 the town suffered its worst flooding in years. Torrential rain caused drains to block as well as an Anglian Water pumping station to break down and this resulted in flash flooding around the town in which 90 properties were flooded up to 5 ft.[7]

The town was badly affected by the North Sea flood of 1953. On 9 November 2007 the town braced itself for more flooding as a result of a tidal surge and high tides but disaster was avoided and only a small area was under water.[8]

New Ferry Route

Initially the idea of the outer harbour was sold to the public as a ferry terminal with 150,000 trucks passing through yearly and 10,000 jobs created. Before the deal was signed for the construction of the Outer Harbour Superfast Ferries disappeared from the scene and in a recent press release Managing Director Eddie Freeman of Great Yarmouth Port Company Ltd was reported as saying that he had never promised a ro-ro terminal. This seems to be reinforced by Peter Hardy of Great Yarmouth Borough Council who in the same article was reported as saying people should not get too hung up about a ro-ro terminal.[citation needed]

Sights

The Tollhouse, with dungeons, dates from the late 13th century and is said to be the oldest civic building in Britain. It backs on to the central library.

The Market place is one of the largest in England, and has been operating since the 13th century. It is also home to the town's shopping sector and the famous Yarmouth chip stalls. The smaller area south of the market is used as a performance area for community events and for access to the town's shopping centre,

Panorama of Hall Quay seen from Southtown. This shows the Town Hall and Star Hotel. Historic South Quay continues to the right of the image.

Market Gates. In November 2008, a new section of Market Gates opened, including high street retailers such as Debenhams, New Look and Starbucks.

Great Yarmouth railway station, which serves the town, is the terminus of the Wherry Lines from Norwich. Before the Beeching Axe the town had a number of railway stations and a direct link to London down the east coast. The only remaining signs of these stations is the coach park where Beach Station once was and the A12 relief road which follows the route of the railway down into the embankment from Breydon Bridge.

Yarmouth has two piers, Britannia Pier and Wellington Pier. The theatre building on the latter of the two was demolished in 2005 and is currently being rebuilt as a family entertainment centre. Britannia Pier is home to the Britannia Theatre which during the summer months features well known acts including; Jim Davidson, Jethro, Basil Brush, Cannon and Ball, Chubby Brown, Chuckle Brothers and The Searchers. The theatre is one of a few end of the pier theatres left in England.

The Grade II listed Winter Gardens building sits next to the Wellington Pier. The cast iron framed glass structure was shipped by barge from Torquay in 1903. It is said this was done without the loss of a single pane of glass. Over the years, it has been used as ballroom,

Britannia Pier

roller skating rink and beer garden. In the 1990s it was converted into a nightclub by comedian Jim Davidson. Today, The Winter Gardens are used as a family leisure venue, although its future is under threat owing to the cost of repairing the aging framework. During the winter of 2005 there were worries that building might collapse, and during high winds it was often closed.

Great Yarmouth's seafront, known as "The Golden Mile" attracts hundreds of thousands of visitors each year to its sandy beaches, Joyland, outdoor attractions and amusement arcade. Great Yarmouth's Marine Parade has 12

Amusement Arcades located within 2 square miles (**unknown operator: u'strong'** km^2), including: Atlantis, The Flamingo, Circus Circus, The Golden Nugget, The Mint, Leisureland, The Majestic, The Silver Slipper, The Showboat, Magic City, Quicksilver and The Gold Rush, opened in 2007.

The South Denes area is home to the Grade I listed Norfolk Naval Pillar, known locally as Nelson's Monument or Nelson's Column. This tribute to Admiral Lord Horatio Nelson was completed in 1819, 24 years before the completion of Nelson's Column in London. The monument, designed by William Wilkins, shows Britannia standing atop a globe holding an olive branch in her right hand and a trident in her left.There is a popular assumption in the town that the statue of Britannia was supposed to face out to sea but now faces inland due to a mistake during construction, although it is thought she is meant to face Nelson's birthplace at Burnham Thorpe. The monument was originally planned to mark Nelson's victory at the Battle of the Nile, but fund-raising was not completed until after his death and it was instead dedicated to England's greatest Naval hero. It is currently surrounded by an industrial estate but plans are in place for the improvement of the area. The Norfolk Nelson Museum on South Quay houses the Ben Burgess collection of Nelson Memorabilia and is the only dedicated Nelson museum in Britain other than one in Monmouth. Its several galleries look at Nelson's life and personality as well as what life was like for the men who sailed under him.

Charles Dickens used Yarmouth as a key location in his novel *David Copperfield*. The author stayed at the Royal Hotel on the Marine parade while writing David Copperfield. Anna Sewell (1820–1878), the author of *Black Beauty*, was born in a 17th century house in Church Plain. The house is currently being used as a restaurant after being renovated in 2007.

The Time and Tide Museum on Blackfriars Road which is managed by Norfolk Museums Service was nominated in the UK Museums Awards in 2005. It was built as part of the regeneration of the south of the town in 2003. Its location in an old herring smokery harks back to the town's status as a major fishing port. Sections of the historic town wall are located outside the museum.

Small boat at the Time and Tide museum

The Maritime Heritage East partnership, based at the award winning Time and Tide Museum aims to raise the profile of maritime heritage and museum collections.

Wildlife

The Yarmouth area is home to a number of rare and unusual species. The area between the piers is home to one of the largest roosts of Mediterranean Gulls in the UK. Breydon Water, just behind the town, is a major wader and waterfowl site, with winter roosts of over 100,000 birds. This and the surrounding Halvergate Marshes are specially protected, and the majority of the area is now owned by conservation organisations, principally the RSPB.

The North Denes area of the beach is an SSSI due to its dune plants, and is home to numbers of Skylarks and Meadow Pipits. It also hosts one of the largest Little Tern colonies in the Uk each summer, as well as a small colony of Greyling butterflies. Other butterflies found here include Small Copper and Common Blue.

The near-by cemetery is renowned as a temporary roost for spring and autumn migrants. Redstart and Pied Flycatcher are often seen here during migration. It has also been the site for the first records of a number of rare insects, blown in from the continent.

Central Beach close to the Jetty

Grey Seal and Common Seal are frequently seen off-shore, as are sea-birds such as Gannet, Little Auk, Common Scoter, Razorbill and Guillemot.

Sports and leisure

The main local football club is Great Yarmouth Town, also known as the Bloaters, who play in the Eastern Counties League. Currently managed by Mike Derbyshire, their ground is at Wellesley Recreation Ground (named after Sir Arthur Wellesley). They have a strong East Anglian rivalry with Gorleston. The club has enjoyed limited success in the past. Local football clubs are served by the Great Yarmouth and District League.

Yarmouth has a horse-racing track which features a chute allowing races of one mile (1.6 km) on the straight.

Speedway racing was staged in Great Yarmouth before and after the Second World War. The meetings were staged at the greyhound stadium in Caister Road. The post war team were known as the Yarmouth Bloaters (after the smoked fish). Banger and Stock car racing is also staged at this stadium.

The main Leisure Centre is the Marina Centre. Built in 1981 the centre has a large swimming pool, conference facilities and live entertainment including their famous Summer Pantomimes and Summer Variety Shows produced by local entertainers Hanton & Dean. The centre is run by the Great Yarmouth Sport and leisure Trust. The Trust was set up in April 2006 to run the building as a charitable non profit making organisation.

At the beginning of the 2008 summer season a worlds first Segway Grand Prix was opened at the Pleasure Beach gardens.

Pop Beach

In 2003 and 2004 T4 hosted their T4 on the Beach music festival in the town. It attracted around 20,000 people to the town but was moved to Weston-super-Mare in 2005.[9]

Transport

Rail

Great Yarmouth is connected to Norwich by the Wherry Lines from Great Yarmouth railway station, it is served by an hourly service provided by Greater Anglia via Acle or, less frequently, via Reedham.[10] It is the only remaining station of the three once in the town.

The town was adversely affected by the Beeching Axe. There were previously four railway lines coming into the town; from the north the line came down the coast from Melton Constable into the terminus at Beach station. From the south-west came the line from London via Beccles and from the south the line from Lowestoft Central via Hopton and Gorleston; both of which terminated at South Town station.[11] The only existing terminal is Vauxhall station which is now simply referred to as 'Great Yarmouth station'. Whereas once the town had 17 stations within the borough limits it now only has one.

Bus

The bus station in Great Yarmouth is the major hub for local routes and is located under Market Gates Shopping Centre. The X1 route operated by First Eastern Counties provides a long-distance link between Peterborough and Lowestoft via Norwich and King's Lynn. Other local bus services link the suburban areas of Martham, Hemsby, Gorleston, Bradwell and Belton/ These are mostly operated by First Eastern Counties and services to towns further afield such as Beccles, Southwold, Acle and Diss are mostly operated by Anglian Bus.

Port and River

The River Yare cuts off Great Yarmouth from other areas of the borough such as Gorleston and Southtown and so the town's two bridges have become major transport links. Originally Haven Bridge had been the only link over the river but in the late 1980s Breydon Bridge was built to take the A12 over Breydon Water replacing the old railway bridge; the Breydon Viaduct.[12] both bridges can open to allow river traffic underneath which can result in traffic tailbacks. The phrase "the bridge was up" has become synonymous in the town with being late for appointments.

Haven Bridge; one of the two main links to the town in the upright position to allow boats to pass underneath.

A ferry had been provided between the southerly tip of Great Yarmouth and Gorleston, it provided a way of getting between the factories on South Denes and the mostly residential areas of Gorleston without taking a long detour over the bridge. However; due to increased running costs and the decline of industrial activity in the town it was closed in the early 1990s.[13]

Since 2006, the restored pleasure steamer the *Southern Belle* has provided regular river excursions from the town's Haven Bridge. Built in 1925 for the Earl of Mount Edgcumbe. Today, she is owned by the Great Yarmouth and Gorleston Steam Packet Company Limited.[14]

Construction work on the Outer Harbour began in June 2007 and was completed by 2009, it is a deep-water harbour on the North Sea. Originally there were plans for a ro-ro ferry link with IJmuiden which has failed to materialise, similarly despite the installation of two large cranes in 2009 plans for a container terminal have also been scrapped.[15]

Lifeboat station

There has been a lifeboat in Great Yarmouth since at least 1802. The early boats were privately operated until 1857 when the RNLI took over.[16] The lifeboat station is located on Riverside Road (52°34′32″N 1°43′55″E) from where are operated the Trent class lifeboat *Samarbeta* and the B class (inshore) lifeboat *Seahorse IV*.[17]

The view from the top of the Atlantis Tower showing the Golden Mile and, in the distance, the Outer Harbour

Road

The A12 terminates in the town as do the A143 and the A47 roads. The relief road was built along the path of the old railway to carry the A12 onwards to Lowestoft and London. Congestion is a major problem in the town and roundabouts, junctions and bridges can become gridlocked at rush hour.

Proposed third river crossing

Plans have been put forward for the construction of a third river crossing in Great Yarmouth which would link northern Gorleston with the South Denes and the Outer Harbour avoiding the congested town centre. A public consultation took place in mid-2009 over the four possible proposals but as of late-2010 the plans are stalled by a lack of funding and the closure of the container terminal.[18]

Air

The North Denes Heliport is located to the north of the town and currently operated by CHC Helicopter but it is planned to close in 2011 with operations being moved to Norwich International Airport.[19]

First Responders

There is an East of England Ambulance Service First Responder group setup to cover the Great Yarmouth area. A First Responder scheme is made up of a group of volunteers who within the community in which they live or work, have been trained to attend emergency 999 calls by the NHS Ambulance Service.[20] Responders are members of the community who are trained to use an Automated External Defibrillator, Oxygen and other lifesaving equipment to assist ambulance crews. Community First Responders are not a substitute for the Ambulance Service. But, as they are based within Great Yarmouth, they are able to attend the scene of an emergency call in a very short time; often within the first few minutes and, in the majority of incidents, they will be first on scene. Community First Responders are able to administer vital, life saving first aid treatment before the arrival of an Ambulance crew. Their early intervention further increases the patient's chance of survival or may just provide the simple reassurance that the ambulance crew are on their way. Great Yarmouth First Responders themselves are not paid, being a First Responder is completely voluntary. However the Community First Responder schemes are not funded by the East of England Ambulance Service and rely solely on help from the publics donations.

Notable residents

- Vice-Admiral Horatio Nelson (1758–1805)
- James Beeching (1788–1858) was a shipbuilder in the town, and his firm continued after his death well into the 20th century.
- Oscar winning cinematographer Jack Cardiff (1914–2009)
- English physician and writer Dr Thomas Girdlestone (1758–1822)
- Captain George William Manby (1765–1854), inventor of marine lifesaving equipment
- William Hovell (1786–1875), explorer of Australia was born in Great Yarmouth.
- Sir James Paget (1814–99), Victorian Surgeon who had the James Paget Hospital named in his honour.
- Anna Sewell (1820–78), author of *Black Beauty*
- Leading English Independent Minister William Bridge (c1600-70)
- Alternative rock band Catherine Wheel were from Great Yarmouth
- Rebecca Nurse (1621-1692) celebrated victim of the Salem Witch Trials, was born in Great Yarmouth.
- Jason Statham (Born 12 September 1967, age 44), Actor, lived in Great Yarmouth during the majority of his childhood and attended the local Grammar School.
- Ed Graham (born 20 February 1977 in Great Yarmouth, England), Drummer with the rock band The Darkness (band).

Twinning

Great Yarmouth is twinned with:

- ▌ ▌ Rambouillet, France

See also

- Great Yarmouth Pleasure Beach
- St Nicholas Church, Great Yarmouth
- Lydia Eva the last surviving steam drifter of the Great Yarmouth herring fishing fleet

References

[1] Ordnance Survey (2005). *OS Explorer Map OL40 - The Broads*. ISBN 0-319-23769-9.

[2] "Town's last fishing boat fights tide and time" (http://www.telegraph.co.uk/news/uknews/1575561/Towns-last-fishing-boat-fights-tide-and-time.html). *The Daily Telegraph*. 14 January 2008. .

[3] Office for National Statistics & Norfolk County Council (2001). *Census population and household counts for unparished urban areas and all parishes* (http://www.norfolk.gov.uk/consumption/groups/public/documents/general_resources/ncc017867.xls). Retrieved 2 December 2005.

[4] Nuttall, P Austin (1840). *A classical and archæological dictionary of the manners, customs, laws, institutions, arts, etc. of the celebrated nations of antiquity, and of the middle ages* (http://www.google.co.uk/books?id=V-gDAAAAQAAJ&pg=PA555&dq=Yarmouth+pasties&as_brr=1). London. p. 555. .

[5] Daniel Defoe, *A tour thro' the whole island of Great Britain, divided into circuits or journies* (1724), Letter 1, Pt 3. Defoe's several descriptions may be accessed on the Vision of Britain website (http://www.visionofbritain.org.uk/index.jsp)

[6] "The Fall of Yarmouth Road" (http://www.jeron.je/anglia/learn/sec/history/yarmouth/page03.htm). AngliaCampus. . Retrieved 11 October 2009.

[7] "England | Norfolk | Homes under water in flash floods" (http://news.bbc.co.uk/1/hi/england/norfolk/5378080.stm). BBC News. 25 September 2006. . Retrieved 29 January 2010.

[8] Turner, Andrew (9 November 2007). "England | Norfolk | Residents ride out storm surge" (http://news.bbc.co.uk/1/hi/england/norfolk/7086419.stm). BBC News. . Retrieved 29 January 2010.

[9] Event Info (http://www.channel4.com/entertainment/t4/microsites/P/popbeach/eventinfo/eventinfo.html) Pop Beach Website; Retrieved 20 November 2010]

[10] National Express East Anglia Timetables (http://www.nationalexpresseastanglia.com/travel_information/train_timetables/current_timetable__2/(station)/GYM). Retrieved 20 November 2010.

[11] New Adlestrop Railway Atlas (http://www.systemed.net/atlas/) Retrieved 20 November 2010

[12] Building the Breydon Bridge, June 1985 (http://www.ourgreatyarmouth.org.uk/page_id__454_path__0p44p63p.aspx) Our Great Yarmouth, Retrieved 20 November 2010

[13] Great Yarmouth Ferry Crossings (http://www.ourgreatyarmouth.org.uk/page_id__472_path__0p44p75p.aspx) Our Great Yarmouth; Retrieved 20 November 2010

[14] The Southern Belle (http://www.angliawebsites.com/southern-belle-ferry-great-yarmouth.html). Retrieved 29 October 2009.

[15] Stephen Pullinger (10 November 2010). "Great Yarmouth outer harbour's £7m cranes to go" (http://www.edp24.co.uk/business/great_yarmouth_outer_harbour_s_7m_cranes_to_go_1_722070). Eastern Daily Press. . Retrieved 20 November 2010.

[16] RNLI history of Great Yarmouth & Gorleston lifeboat station (http://www.rnli.org.uk/rnli_near_you/east/stations/GreatYarmouthandGorlestonNorfolk/history)

[17] "Great Yarmouth and Gorleston Fleet" (http://www.rnli.org.uk/rnli_near_you/east/stations/GreatYarmouthandGorlestonNorfolk/fleet). RNLI. . Retrieved 29 January 2010.

[18] "Cash concerns over third crossing" (http://www.eastcoastlive.co.uk/news/info.php?news=Cash+concerns+over+third+crossing&refnum=2188) East Coast Live; Retrieved 20 November 2010

[19] "Town's heliport to close next year" (http://www.eastcoastlive.co.uk/news/info.php?news=Towns+heliport+to+close+next+year&refnum=2279) East Coast Live; Retrieved 20 November 2010]

[20] "Great Yarmouth First Responders" (http://www.gyresponders.org). Great Yarmouth First Responders. . Retrieved 22 June 2010.

Further reading

- Ferry, Kathryn (2009) "'The maker of modern Yarmouth': J. W. Cockrill", in: Ferry, Kathryn, ed. *Powerhouses of Provincial Architecture, 1837-1914*. London: Victorian Society; pp. 45-58

External links

- Official website of Great Yarmouth Borough Council (http://www.great-yarmouth.gov.uk)

Central_London

Central London is the innermost part of London, England. There is no official or commonly accepted definition of its area, but its characteristics are understood to include a high density built environment, high land values, an elevated daytime population and a concentration of regionally, nationally and internationally significant organisations and facilities. From time to time, and for a variety of purposes, a number of definitions have been used to define its scope.

OpenStreetMap of central London.

Road distances to London are traditionally measured from a central point at Charing Cross, which is marked by the statue of King Charles I at the junction of the Strand, Whitehall and Cockspur Street, just south of Trafalgar Square.[1]

Characteristics

> The central area is distinguished, according to the Royal Commission, by the inclusion within its boundaries of Parliament and the Royal Palaces, the headquarters of Government, the Law Courts, the head offices of a very large number of commercial and industrial firms, as well as institutions of great influence in the intellectual life of the nation such as the British Museum, the National Gallery, the Tate Gallery, the University of London, the headquarters of the national ballet and opera, together with the headquarters of many national associations, the great professions, the trade unions, the trade associations, social service societies, as well as shopping centres and centres of entertainment which attract people from the whole of Greater London and farther afield.
>
> In many other respects the central area differs from areas farther out in London. The rateable value of the central area is exceptionally high. Its day population is very much larger than its night population. Its traffic problems reach an intensity not encountered anywhere else in the Metropolis or in any provincial city, and the enormous office developments which have taken place recently constitute a totally new phenomenon.
>
> — Parliamentary Debates, House of Commons, 24 January 1963., Eric Lubbock

Definitions

London Plan

The London Plan includes a central activities zone policy area. This comprises the City of London, most of Westminster and the inner parts of Camden, Islington, Hackney, Tower Hamlets, Southwark, Lambeth and Kensington and Chelsea.[2] It is described as "a unique cluster of vitally important activities including central government offices, headquarters and embassies, the largest concentration of London's financial and business services sector and the offices of trade, professional bodies, institutions, associations, communications, publishing, advertising and the media".[3]

For strategic planning, from 2004 to 2008, the London Plan included a sub-region called Central London comprising Camden, Islington, Kensington and Chelsea, Lambeth, Southwark, Wandsworth and Westminster.[4] It had a 2001 population of 1,525,000. The sub-region was replaced in 2008 with a new structure which amalgamated inner and outer boroughs together.

Census

The 1901 census defined Central London as the City of London and the metropolitan boroughs of Bermondsey, Bethnal Green, Finsbury, Holborn, Shoreditch, Southwark, Stepney, St Marylebone and Westminster.[5]

1959–1963 proposals for a Central London borough

During the Herbert Commission and the subsequent passage of the London Government Bill, three attempts were made to define an area that would form a central London borough. The first two were detailed in the 1959 Memorandum of Evidence of the Greater London Group of the London School of Economics.

'Scheme A' envisaged a central London borough, one of 25, consisting of the City of London, Westminster, Holborn, Finsbury and the inner parts of St Marylebone, St Pancras, Chelsea, Southwark and Lambeth. The boundary deviated from existing lines in order to include all central London railway stations, the Tower of London and the museums, such that it included small parts of Kensington, Shoreditch, Stepney and Bermondsey. It had an estimated population of 350,000 and occupied 7000 acres (**unknown operator: u'strong'** km^2).[6]

'Scheme B' delineated central London, as one of 7 boroughs, including most of the City of London, the whole of Finsbury and Holborn, most of Westminster and Southwark, parts of St Pancras, St Marylebone, Paddington and a small part of Kensington. The area had an estimated population of 400,000 and occupied 8000 acres (**unknown operator: u'strong'** km^2).[6]

During the passage of the London Government Bill an amendment was put forward to create a central borough corresponding to the definition used at the 1961 census. It consisted of the City of London, all of Westminster, Holborn and Finsbury; and the inner parts of Shoreditch, Stepney, Bermondsey, Southwark, Lambeth, Chelsea, Kensington, Paddington, St Marylebone and St Pancras. The population was estimated to be 270,000.[7]

Panorama of central London as seen from the London Eye

References

[1] "OS MapZone - From where, exactly, are distances from London measured?" (http://mapzone.ordnancesurvey.co.uk/mapzone/ didyouknow/distance/q_15_21.html). Ordnance Survey. . Retrieved 2010-03-10.

[2] Mayor of London (2008). "Central activities zone" (http://www.london.gov.uk/thelondonplan/images/maps-diagrams/jpg/map-5g-1. jpg). *London Plan*. Greater London Authority. . Retrieved 10 March 2010.

[3] Mayor of London (2008). "Central activities zone policies" (http://www.london.gov.uk/thelondonplan/images/maps-diagrams/jpg/ map-5g-1.jpg). *London Plan*. Greater London Authority. . Retrieved 10 March 2010.

[4] Mayor of London (February 2004). "The London Plan: Chapter 5" (http://www.london.gov.uk/archive/mayor/strategies/sds/ london_plan/lon_plan_5.pdf). Greater London Authority. .

[5] "1901 Census of England and Wales, General Report with Appendices (1904 CVIII (Cd. 2174) 1)" (http://www.visionofbritain.org.uk/ text/chap_page.jsp?t_id=SRC_P&c_id=4&cpub_id=EW1901GEN&show=DB). . Retrieved 10 March 2010.

[6] Greater London Group (July, 1959). *Memorandum of Evidence to The Royal Commission on Local Government in Greater London*. London School of Economics.

[7] *Parliamentary Debates*, House of Commons, 24 January 1963.

Dereham

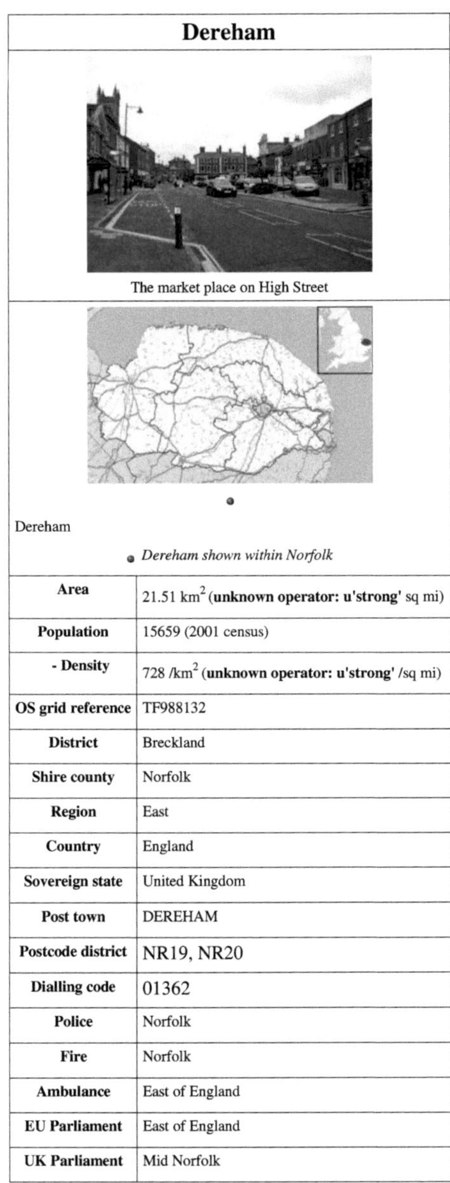

Dereham

The market place on High Street

Dereham

Dereham shown within Norfolk

Area	21.51 km^2 (**unknown operator: u'strong'** sq mi)
Population	15659 (2001 census)
- Density	728 /km^2 (**unknown operator: u'strong'** /sq mi)
OS grid reference	TF988132
District	Breckland
Shire county	Norfolk
Region	East
Country	England
Sovereign state	United Kingdom
Post town	DEREHAM
Postcode district	NR19, NR20
Dialling code	01362
Police	Norfolk
Fire	Norfolk
Ambulance	East of England
EU Parliament	East of England
UK Parliament	Mid Norfolk

Dereham, also known as **East Dereham**, is a town and civil parish in the English county of Norfolk. It is situated on the A47 road, some 15 miles (25 km) west of the city of Norwich and 25 miles (40 km) east of King's Lynn. The civil parish has an area of 21.51 km^2 (**unknown operator: u'strong'** sq mi) and in the 2001 census had a population of 15,659 in 6,941 households. For the purposes of local government, Dereham falls within, and is the centre of administration for, the district of Breckland.[1]

Since 1983 Dereham has been twinned with the town of Rüthen in North Rhine-Westphalia, Germany. It is also twinned with Caudebec-lès-Elbeuf - France

History

A map of Dereham from 1946

Early history

It is believed that Dereham's name derives from a deer park that existed in the area, although it is known that the town pre-dates the Saxon era.[2] Saint Wihtburh, the youngest daughter of Anna, King of the East Angles, founded a monastery there in the seventh century after seeing a vision of the Virgin Mary; the monastery is mentioned by Bede, but little further is known of it.

Many of the town's ancient buildings were destroyed in the serious fires that took place in 1581 and 1659. Notable buildings that survived the fire include the Church of Saint Nicholas' and the nearby Bishop Bonner's cottage.

Railways

The railway arrived in Dereham when a single track line to Wymondham opened in 1847. In 1848 a second line, to King's Lynn was opened. In 1849 a line from Dereham to Fakenham was opened, this line being extended to the coastal town of Wells-On-Sea by 1857. In 1862 the town's railways became part of the Great Eastern Railway. The town had its own railway depot and a large complex of sidings, serving local industry. In 1882 the line between Dereham and Wymondham was doubled, to allow for the increasing levels of traffic.

In 1964 passenger services between Dereham and Wells were withdrawn, and the track between Fakenham and Wells was lifted soon after. In 1965 the line from Dereham to Wymondham was returned to single track, with a passing loop at Hardingham. The line to King's Lynn was closed in 1968, and the last passenger train on the Dereham to Wymondham line ran in 1969 although the railway remained open for freight until 1989.

Dereham labels itself "The Heart of Norfolk" due to its central location in the county, the Tesco car park being cited as the exact centre.[3] In the spring of 1978, the "Heart" was given the seven-mile £5m part-dual-carriageway A47 bypass. A section of this road, between *Scarning* and *Wendling* was built along the former railway line towards Swaffham and King's Lynn. This section of railway had been used as a location for the filming of Dad's Army, where Captain Mainwaring is left dangling from a railway bridge after a flight on a barrage balloon.

The railway between Dereham and Wymondham has been preserved, and is now operated as a tourist line by the Mid-Norfolk Railway Preservation Trust. This charitable company also owns the line north towards County School railway station and aims to eventually relay the line to Fakenham. The town should not be confused with the Norfolk village of West Dereham, which lies about 25 miles (40 km) away.

Saint Withburga

The town lies on the site of a monastery founded by Saint Withburga in the seventh century. A holy well at the western end of St Nicholas' Church supposedly began to flow when her body was stolen from the town by monks from Ely, who took the remains back to their town.

In the 18th century an attempt was made to turn Dereham into a new Buxton or Bath by building a bath house over Withburga's Well. It was described at the time as a hideous building of brick and plaster and was never popular. In 1880 the local vicar, Reverend Benjamin Armstrong obtained permission to pull the building down. The spring was then protected by iron railings, but fell out of use and became choked with weeds. Since 1950, however, it has been kept clear of weeds—although the railings still prevent access to the waters. Close examination of the Withburga story will cast doubt on Dereham being the location of the Saint's abode and resting place. The legend states that monks from Ely came 'up the river' at night and stole her body, taking it back to Ely to rest with her sisters, who were already considered saints. A look at a map will prove this to be an impossibility as there is no river connecting Ely with East Dereham, although it is possible to navigate a river from Ely to West Dereham. Until proved otherwise, Dereham continues to be considered the site of Withburga's home and violated grave.

Industry and employment

Dereham was the home to the "Jentique" furniture factory which made boxes for both instruments and bombs during the 2nd World War. The town was also the home to the Metamec clock factory. "Hobbies of Dereham" produced plans, kits and tools, including their famous treadle fretwork saws, for making wooden models and toys which were popular in the days before moulded plastic. At one point Hobbies owned 10 shops in prestige locations all over the UK. In the early 1960s the firm was taken over by Great Universal Stores, who sold the shops and closed the business. However, due to a shrewd management purchase of the "old traditional" parts of the firm, Hobbies rose again, limiting itself to a role as specialist model makers shop. After nearly 40 years of its new lease of life the firm closed in 2009. Cranes of

One of several maltings in the area, also looking at the Railway line which closed in 1989 to freight (now MNR)

Dereham and its successor the Fruehauf trailer company was a major employer in the town for many decades. Cranes built nearly all the giant trailers (100 tons plus) that carried equipment such as transformers at slow speeds across the UK, usually in the livery of Wynns or Pickfords. The launch of a new trailer was treated rather like that of a ship with lots of people coming out to see the leviathan move through the narrow streets of the town towards the A47. The town also had several large maltings. Almost all this large scale industry has drifted away since the 1980s.

Sport & Leisure

Dereham has a Non-League football club Dereham Town F.C. who play at Aldiss Park.

Youth and community provision

Schools

Nursery and pre-school

- Toftwood Nursery Pre-School
- Scarning Pre-School

Infant and junior schools

- Scarning Primary school
- Grove House Nursery and Infant Community School
- Dereham CE VA Infant School and Nursery
- Dereham St. Nicholas Junior School
- King's Park Infant School
- Toftwood Infant School

Secondary schools

- Neatherd High School
- Northgate High School

Sixth form college

- Dereham Sixth Form College

East Dereham Windmill

Scouts

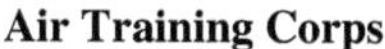

Dereham has two active Scout Groups, both of which are part of The
Scout Association. 1st Dereham is notable as one of the earliest Groups set up in the world, having been formed in 1908. In the past there was also a 3rd Dereham Scout Group.

Air Training Corps

The town is the home of 1249 Squadron, Air Training Corps, who parade at the Cadet Centre on Norwich Road.

Army Cadet Force

The Army Cadets also parade at the Cadet Centre on Norwich Road.

Attractions

Notable buildings in the town include the pargetted Bishop Bonners Cottage, built in 1502, the Norman parish church, a windmill and a large mushroom-shaped water tower. The Gressenhall Museum of Rural Life is nearby. The town also hosts the headquarters of the Mid-Norfolk Railway, which runs trains over an 11.5-mile railway to Wymondham, as well as owning the line 6 miles north to North Elmham and County School Station.

Notable people

Famous people from the town include novelist Brian Aldiss, author George Borrow, footballer Harry Cripps, the antiquarian Sir John Fenn, Lady Ellenor Fenn, architect George Skipper, singer Beth Orton and scientist William Hyde Wollaston. William Cowper, poet, died in Dereham, and is buried in St Nicholas's Church, where there is a commemorative stained glass window. The Oldhall family held the manor in the fourteenth and fifteenth centuries: notable members of the family included Sir William Oldhall, Speaker of the House of Commons and his brother Edmund Oldhall, Bishop of Meath.

Cowper Church Sunday School, Dereham.

References

[1] Office for National Statistics & Norfolk County Council (2001). *Census population and household counts for unparished urban areas and all parishes* (http://www.norfolk.gov.uk/consumption/groups/public/documents/general_resources/ncc017867.xls). Retrieved December 2, 2005.

[2] http://www.derehamtimes.co.uk/content/derehamtimes/content/Community/DerehamHistory.aspx Dereham Times, About Dereham

[3] http://www.edp24.co.uk/Content/Postcard_From/dereham.asp Eastern Daily Press, Postcard from Dereham

External links

- Dereham Town Council website (http://derehamtc.norfolkparishes.gov.uk/)
- Information from Genuki Norfolk (http://www.origins.org.uk/genuki/NFK/places/d/dereham_east/) on Dereham.

Swaffham

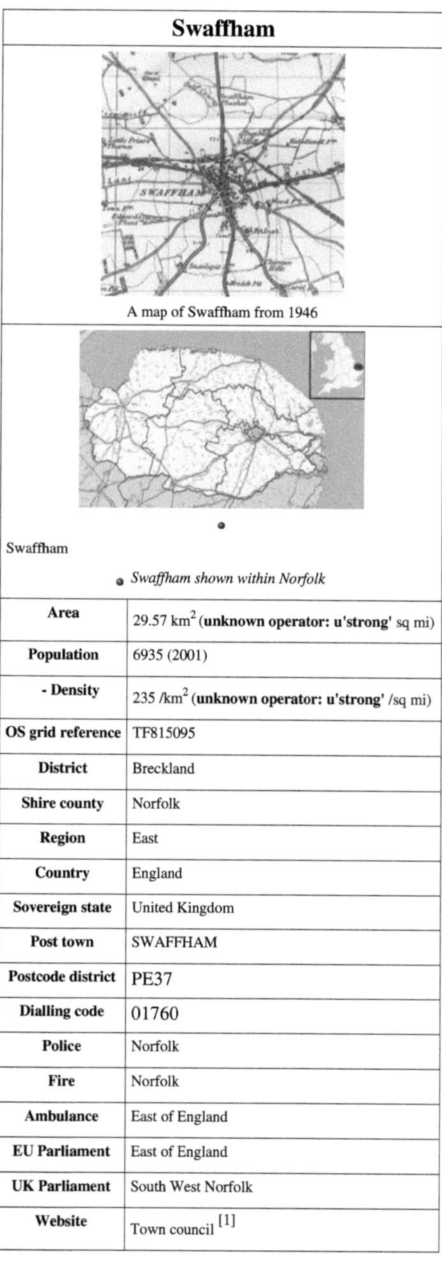

<table>
<tr><th colspan="2" align="center">Swaffham</th></tr>
<tr><td colspan="2" align="center">A map of Swaffham from 1946</td></tr>
<tr><td colspan="2" align="center">Swaffham

Swaffham shown within Norfolk</td></tr>
<tr><td>Area</td><td>29.57 km^2 (unknown operator: u'strong' sq mi)</td></tr>
<tr><td>Population</td><td>6935 (2001)</td></tr>
<tr><td>- Density</td><td>235 /km^2 (unknown operator: u'strong' /sq mi)</td></tr>
<tr><td>OS grid reference</td><td>TF815095</td></tr>
<tr><td>District</td><td>Breckland</td></tr>
<tr><td>Shire county</td><td>Norfolk</td></tr>
<tr><td>Region</td><td>East</td></tr>
<tr><td>Country</td><td>England</td></tr>
<tr><td>Sovereign state</td><td>United Kingdom</td></tr>
<tr><td>Post town</td><td>SWAFFHAM</td></tr>
<tr><td>Postcode district</td><td>PE37</td></tr>
<tr><td>Dialling code</td><td>01760</td></tr>
<tr><td>Police</td><td>Norfolk</td></tr>
<tr><td>Fire</td><td>Norfolk</td></tr>
<tr><td>Ambulance</td><td>East of England</td></tr>
<tr><td>EU Parliament</td><td>East of England</td></tr>
<tr><td>UK Parliament</td><td>South West Norfolk</td></tr>
<tr><td>Website</td><td>Town council [1]</td></tr>
</table>

Swaffham is a market town and civil parish in the English county of Norfolk. The town is situated 20 km (**unknown operator: u'strong'** mi) east of King's Lynn and 50 km (**unknown operator: u'strong'** mi) west of Norwich.

The civil parish has an area of 29.57 km^2 (**unknown operator: u'strong'** sq mi) and in the 2001 census had a population of 6,935 in 3,130 households. For the purposes of local government, the parish falls within the district of Breckland.[2]

History

The Buttercross Swaffham market place.

Its name came from Old English *Swæfa hām* = "the homestead of the Swabians"; some of them presumably came with the Angles and Saxons. In the Domesday Book three lords were associated with Swaffham: Walter Giffard, with the largest manor; his tenant Hugh Bolebec, who held all of the Giffard land there; and Aubrey de Vere I, who held a smaller manor at Swaffham which the Domesday jurors said Aubrey had seized without the king's permission.[3] As the Bolebec estates passed into Vere hands through two marriages of Bolebec heiresses to Vere males in the late 12th and early 13th centuries, the two manors were combined and held by the Vere Earls of Oxford for several centuries.

A Benedictine priory for female religious was founded at Swaffham Bolebec between *circa* 1150 and 1163, probably by the Bolebecs.[4] About 8 km to the north of Swaffham can be found the ruins of the formerly important Castle Acre Priory and Castle Acre Castle.

By the 14th and 15th centuries Swaffham had a flourishing sheep and wool industry As a result of this prosperity, the town has a large market place. The Market Cross here was built by George Walpole, 3rd Earl of Orford and presented to the town in 1783.[5] On the top is the statue of Ceres, the Roman goddess of the harvest.

On the west side of Swaffham Market Place are several old buildings which for many years housed the historic Hamond's Grammar School, as a plaque on the wall of the main building explains. The Hamond's Grammar School building latterly came to serve as the sixth form for the Hamond's High School, but that use has since ceased. Harry Carter, the Grammar School's art teacher of the 1960s, was responsible for a great number of the carved village signs that are now found in many of Norfolk's towns and villages, most notably perhaps Swaffham's own sign commemorating the legendary Pedlar of Swaffham,[6] [7] [8] which is in the corner of the market place just opposite the old school's gates.[9] Carter was a distant cousin of the archaeologist and egyptologist Howard Carter[10] who spent much of his childhood in the town.[11]

Until 1968 it had a railway station on the Great Eastern Railway line from King's Lynn. Just after Swaffham, the line split into two, one branch heading south to Thetford, and the other east towards Dereham. The railways were closed as part of the Beeching Axe, through the possibility of rebuilding a direct rail link from Norwich to King's Lynn via Swaffham is occasionally raised.

The Swaffham Museum contains an exhibition on local history and local geology as well as an egyptology room charting the life of Howard Carter.[12]

Ecotech Centre

The Ecotech Centre

Today the town is known for the presence of two large Enercon E-66 wind turbines, and the associated Ecotech Centre.[13] The turbines are owned and operated by Ecotricity, and together generate more than three megawatts.[14] One wind turbine, an Enercon E66/1500 with 1.5 MW generation capacity, 67 metres nacelle height and 66 metres rotor diameter, which was built in 1999,[15] has an observation deck just below the nacelle. These have now been joined now by a further eight turbines at North Pickenham.

The centre hosted the 2008 British BASE jumping championships; contestants jumped from the roof of the observation deck.[16]

Sport and leisure

Swaffham has a Non-League football club Swaffham Town F.C. who play at Shoemakers Lane.

Climate

As with the rest of the British Isles and East Anglia, Swaffham experiences a maritime climate with cool summers and mild winters. The nearest Met Office weather station to provide local climate data is RAF Marham, about 5.5 miles west of the town centre. Temperature extremes in the Swaffham-Marham area range from 34.8 °C (**unknown operator: u'strong'** °F) in August 1990, down to −16.7 °C (**unknown operator: u'strong'** °F) during February 1956.[17] . The highest and lowest temperatures reported in the past decade are 34.6 °C (**unknown operator: u'strong'** °F) during August 2003,[18] and −10.3 °C (**unknown operator: u'strong'** °F) during January 2010.[19] .

Climate data for Marham 21m asl, 1971-2000 (Weather Station 5 miles West of Swaffham)

Month	Jan	Feb	Mar	Apr	May	Jun	Jul	Aug	Sep	Oct	Nov	Dec	Year
Average high °C (°F)	6.6	7.1	10.0	12.2	16.2	19.0	21.7	21.8	18.6	14.3	9.7	7.4	13.8
Average low °C (°F)	0.5	0.6	2.3	4.0	6.9	9.7	11.8	11.8	9.6	6.6	3.2	1.6	5.7
Precipitation mm (inches)	54.7 (2.154)	38.5 (1.516)	49.5 (1.949)	46.8 (1.843)	48.1 (1.894)	55.9 (2.201)	44.1 (1.736)	50.5 (1.988)	54.9 (2.161)	59.8 (2.354)	63.3 (2.492)	55.3 (2.177)	621.3 (24.461)
Mean monthly sunshine hours	53.6	73.2	101.7	150.6	204.3	191.1	202.7	192.8	139.8	109.7	69.0	48.1	1536.6
Source: Met Office[20]													

Kingdom (TV series)

In the summer of 2006, location filming was done in the town for the ITV1 series *Kingdom*, starring Stephen Fry. In *Kingdom* the town is called Market Shipborough. The pub *The Startled Duck* in the TV series is better known as *The Greyhound Inn* in which the Earl of Orford created the first coursing club open to the public in 1776.[21] Kingdom's office is filmed in Oakleigh House, near the town square (formerly the house of the Head Master of Hamond's Grammar School), with the coastal scenes filmed at Wells-next-the-Sea on the north Norfolk coast.

Roads

Swaffham fire station.

The east-west A47 Birmingham to Great Yarmouth road now avoids the town, using a northerly bypass opened in 1981. The A1065 Mildenhall to Fakenham road still passes through the centre of the town on its north-south route, intersecting with the A47 at a grade separated junction north of the town.[22]

Notable people

* Michael Carroll, lottery winner
* Howard Carter, archaeologist who discovered the tomb of Tutankhamun
* Christopher Dawes, author of *Rat Scabies And The Holy Grail*
* Stephen Fry, actor and writer
* W.E. Johns, author of the *"Biggles"* books
* William Methwold (1590–1653), born South Pickenham, East India Company merchant
* Sir Arthur Knyvet Wilson, (1842–1921), First Sea Lord

References

[1] http://www.swaffhamtowncouncil.co.uk

[2] Office for National Statistics & Norfolk County Council (2001). *Census population and household counts for unparished urban areas and all parishes* (http://www.norfolk.gov.uk/consumption/groups/public/documents/general_resources/ncc017867.xls). Retrieved December 2, 2005.

[3] Inquisitio Comitatus Cantabrigiensis, *Victoria County History of Cambridgeshire*, vol. I, p. 403.

[4] D. Knowles and Hadcock, *Medieval Religious Houses*, p. 266.

[5] Ripper, B. (1979) *Ribbons from the Pedlar's Pack* p126 ISBN 0-9506728-0-7

[6] The Pedlar of Swaffham. (http://www.sacred-texts.com/neu/eng/meft/meft21.htm) *More English Fairy Tales by Joseph Jacobs (1894)*. Retrieved on 2007-03-27

[7] The Pedlar of Swaffham. (http://www.oldcity.org.uk/norwich/names/pedlar.php) *Old City – Names and Legends*. Retrieved on 2007-03-27

[8] Animation (http://www.aniboom.com/video/298688/The Pedlar of Swaffham/?ref=/Scoreboard/aniBoom_Awards_2008)

[9] Literary Norfolk (http://www.literarynorfolk.co.uk/swaffham.htm) Retrieved 22 July 2011

[10] Google books (http://books.google.co.uk/books?id=AyK0_weGa20C&pg=PA4&lpg=PA4&dq=harry+carter+swaffham& source=bl&ots=AUg3GLgfW1&sig=fqiA5tZa8UkTHAFpxmqantSiQ-4&hl=en&ei=B0wpTqy5D825hAeZvezICw&sa=X& oi=book_result&ct=result&resnum=8&ved=0CE4Q6AEwBw#v=onepage&q=harry carter swaffham&f=false) Retrieved 22 July 2011

[11] Howard Carter (http://www.atlantisring.com/howard_carter.aspx) Retrieved 22 July 2011

[12] Swaffham Museum (http://www.swaffhammuseum.co.uk/default.shtm) Retrieved 22 July 2011

[13] Ecotech Centre (http://www.ecotech.org.uk/)

[14] Ecotricity. *Swaffham-I* (http://www.ecotricity.co.uk/projects/op_ecotech.html) and *Swaffham-II* (http://www.ecotricity.co.uk/ projects/op_swaffhamII.html). Retrieved February 10, 2006.

[15] The Windpower.net (http://www.thewindpower.net/wind-farm-1403.php)

[16] "Turbine hosts base jumping" (http://news.bbc.co.uk/1/hi/england/7643085.stm). BBC News. 29 September 2008. . Retrieved 26 January 2011.

[17] "Marham temperature extremes" (http://www.edp24.co.uk/news/behind_the_scenes_at_the_marham_weather_station_1_526795). EDP.

[18] "Marham temperature 2003" (http://www.tutiempo.net/en/Climate/Marham/10-08-2003/34820.htm). TuTiempo. .

[19] "Marham temperature 2010" (http://www.tutiempo.net/en/Climate/Marham/06-01-2010/34820.htm). EDP. .

[20] "Marham 1971–2000 climate averages" (http://www.metoffice.gov.uk/climate/uk/averages/19712000/sites/marham.html). Met Office. . Retrieved Nov 10 2011.

[21] History of Greyhounds: 18th and 19th Centuries (http://www.gulfcoastgreyhounds.org/hist-18-19-cent.html)

[22] Ordnance Survey (1999). *OS Explorer Map 236 – King's Lynn, Downham Market & Swaffham*. ISBN 0-319-21867-8.

External links

- Swaffham Town Council (http://www.swaffhamtowncouncil.co.uk/)
- Information from Genuki Norfolk on Swaffham (http://www.origins.org.uk/genuki/NFK/places/s/swaffham/)

Fransham

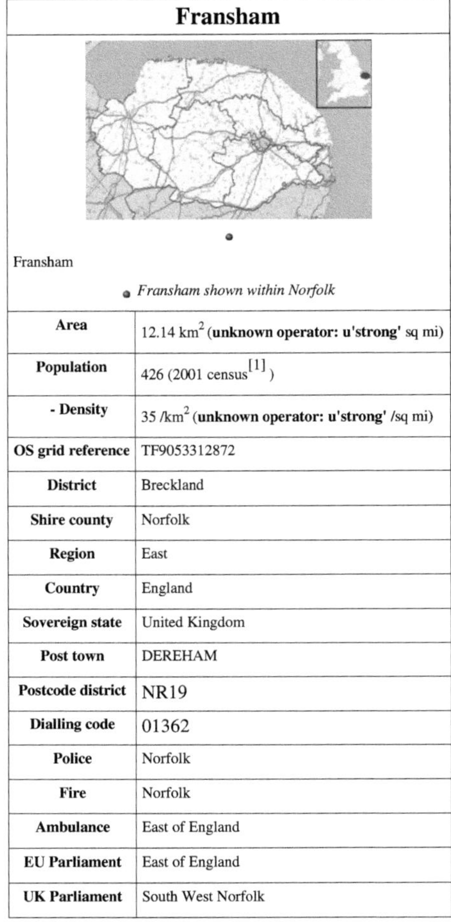

Fransham

Fransham

Fransham shown within Norfolk

Area	12.14 km^2 (**unknown operator: u'strong'** sq mi)
Population	426 (2001 census[1])
- Density	35 /km^2 (**unknown operator: u'strong'** /sq mi)
OS grid reference	TF9053312872
District	Breckland
Shire county	Norfolk
Region	East
Country	England
Sovereign state	United Kingdom
Post town	DEREHAM
Postcode district	NR19
Dialling code	01362
Police	Norfolk
Fire	Norfolk
Ambulance	East of England
EU Parliament	East of England
UK Parliament	South West Norfolk

Fransham is situated in the Breckland District of Norfolk and covers an area of 12.14 km^2 (**unknown operator: u'strong'** sq mi), consisting the villages of Great and Little Fransham, has an estimated population of 430 as of 2007[2]. It lies 6 miles east from Swaffham and 6.5 miles west of Dereham.

The local pub in Little Fransham, the Canary and Linnet, is just off the A47. Great Fransham had a public house called Chequers which is now a private dwelling.

Great Fransham is served by All Saints church[3] and Little Fransham by St.Marys[4] both in the Benefice of Great Dunham[5].

There is a commercially working forge with public demonstrations available.

The unusual Fransham Obelisk raised in the memory of Lord Nelson is actually just in Little Dunham.

Gallery

The Canary and Linnet public house

References

[1] Census population and household counts for unparished urban areas and all parishes (http://www.norfolk.gov.uk/consumption/groups/public/documents/general_resources/ncc017867.xls). Office for National Statistics & Norfolk County Council (2001). Retrieved 20 June 2009.

[2] Breckland Yearbook (http://www.breckland.gov.uk/yearbook_parish_information)

[3] Norfolk churches (http://www.norfolkchurches.co.uk/greatfransham/greatfransham.htm)

[4] Norfolk churches (http://www.norfolkchurches.co.uk/littlefransham/littlefransham.htm)

[5] Benefice of Great Dunham (http://www.norwich.anglican.org/church?benefice=175)

Article Sources and Contributors

Great_Fransham *Source*: http://en.wikipedia.org/w/index.php?title=Great_Fransham *Contributors*: Aboutmovies, ClickRick, Grutness, Lord Cornwallis, Richardguk, The Anome, Wisdom89, 13 anonymous edits

Norfolk *Source*: http://en.wikipedia.org/w/index.php?title=Norfolk *Contributors*: 13eco13, 159753, 28bytes, 30pages, A3RO, AK1712, AWhitwham, Acalamari, Acenditcantu, Adashiel, AdiJapan, Aeusoes1, Agendum, Al Silonov, Albiemax, Amigadave, Amitchell125, Anders.Warga, Andre Engels, Andreakkk, Andrew Gray, AndrewHowse, AndrewMcQ, AndrewRT, Andrewjdavis, Andyblack, Anetode, Angmering, Antandrus, Anwar saadat, Ariel., Aristocrates, Audrey Horne 89, Avono, Barnabypage, Bazonka, Bevo74, Billicus999, Bkell, Bladedfish, Bluemoose, BrownHairedGirl, C1self, CalculatorChad, Canuckian89, Caponer, Castillianthfc, Chanheigeorge, Chaosdruid, Chris j wood, Chris.mciver, Chrisieboy, Chrism, Chtito, Chuunen Baka, Ckatz, Clemo uk, ClickRick, Cnyborg, Colonies Chris, Cometstyles, Conk 9, Conte Giacomo, Costesseyboy, Cry in the night, Cwlhandluke, D Namtar, D6, Danielgibbins, Darklilac, Daveportslade, DavyJonesLocker, Deflective, Dekimasu, Deltron3000, Deor, DerHexer, Dergraaf, Devin122, Dick Bos, DinosaursLoveExistence, Dn9ahx, Don4of4, Donama, Dpaajones, DrKiernan, DuncanHill, Dusimpson, Dysepsion, Eachone, Eajwg02, Edison, Edward, EhJJ, Ehollox, Elmandine, Emeraude, Emmafinn, Epbr123, Esanchez7587, Etacar11, Etm157, Eugene van der Pijll, Exagerative, Felix Folio Secundus, Ferrettini, Fifelfoo, Filemon, Flatterworld, Fluke.magik, Fluoronaut, Francs2000, Freckles290, Fuzjon, Fvasconcellos, Gabbe, Gaius Cornelius, Galwhaa, Giano, Giantsguy06, Gillianatschool, Gimingham, GoUrban, Gogo Dodo, GoingBatty, GraemeLeggett, Grafen, Greenebee, Grezmel, Grinner, Ground Zero, Grstain, Hawkania, Hdt83, HelioSmith, Hemanshu, Henry Flower, Heron, Hide&Reason, Hinders fan, I Hate CAPTCHAS, Ilikeeatingwaffles, Indon, Instamy, Isis, Istabo, Itititititijfjjfjdfjigfvfgnfjg, Jack Phoenix, Jake Nelson, Jamesinderbyshire, Japanese Searobin, Jaydec, Jaymondo77, Jeni, JeremyA, Jerusalemslot, Jevansen, Jfbarlow, Jhamez84, Jickstumps, John, John of Reading, Johnald1337, Joldy, Jool22, Jpeeling, JustAGal, Justin2006, Jza84, K3tts, Kakoui, Kaneaasdf, Keith D, Keith Edkins, Kevyn, Khukri, KingOfTheLynn, KirkEN, Kkkdc, KnightRider, Koine ftw, Kowtoo, Kuru, Kurykh, Kwamikagami, Lam Kin Keung, Langtonboy, Le Creillois, Leszek Jańczuk, Lightmouse, Lilac Soul, LindsayH, Lord Cornwallis, Loren Rosen, Lsm10100, Luwilt, Luís Felipe Braga, MER-C, MJO, MOTORAL1987, MRSC, MSWE1982, Mahahahaneapneap, Malke 2010, Mangwanani, MarmadukePercy, Martarius, Mattis, Max Naylor, Mboverload, McSly, Menchi, Mholland, Mic, Michael Hardy, Minemyfaceoff, Mjh ca, Mmcornick, Mndkingslynn, Modulatum, Mog1122, Morwen, Mossymere, Mr. Stradivarius, Mrshlick, Mugginsx, Multichill, Munderman, NakanoHito, Nasnema, NawlinWiki, Neutrality, NewEnglandYankee, Nick1nildram, Nihiltres, Nilfanion, Nk, Nono64, Norf star, Norm, Northmetpit, Norwikian, Noyder, Ohconfucius, OldSpot61, Oliver Pereira, Oneblackline, Opal-kadett, Orioane, Our Phellap, Owain, OwenBlacker, PDMB, Palmiped, Panickybee, Paul-L, Paulrumsby, Pcpcpc, Peanutashbourne, Pgk, Philip Trueman, Phoe, Pigman, Pjoef, Plucas58, Positivematt, Proclaimer, QuiteUnusual, RF75, RaseaC, Ratarsed, Rebelyell2, Reedjr5746, Renata, Rhillman, Rich Farmbrough, Richhoncho, Ricky81682, Rjwilmsi, Roaringboy, RobertG, RobertWalden, SE7, SMC, SP-KP, SUFCboy, Saga City, Schmiteye, Shirt58, Shrs105, Silverhorse, Simple Bob, Siroxo, Sjakkalle, Sjc, Sjorford, Sky Attacker, Smills1994, Solipsist, Spellcast, Starrycupz, Stavros1, Steinsky, Stemonitis, Stephen, Stephenb, SteveDay, Stooks, Sue Wallace, SunCreator, TBloemink, TRBP, Tagilbert, Tahari29, Tassedethe, Tatrgel, Terri G, Terrifying Angel, Teutonic Tamer, Tharnton345, The Anome, The JPS, The Master of Mayhem, The Norfolk Mead Hotel, The Thing That Should Not Be, TigerShark, Tim Brook, Timrollpickering, Tobby72, TodorBozhinov, Tom Pippens, TourNorfolk, Tresiden, Trontonian, Tsange, Tzartzam, U+003F, Uncle Dick, Upmeath, Utcursch, Valmoer, Vedek Dukat, Vegaswikian, Veinor, WOSlinker, Waggers, Weakopedia, Welsh, WereSpielChequers, Wereon, WhiteKongMan, Whosgotbucket2, William Avery, Wimt, Woodgreener, Xn4, YoterMimeni, Zburh, Zedh, Zzuuzz, Þjóðólfr, ילקורב, , , 445 anonymous edits

King's_Lynn *Source*: http://en.wikipedia.org/w/index.php?title=King%27s_Lynn *Contributors*: Abcrookie, Acalamari, Achangeisasgoodasa, Adam Carr, Adambro, Ahoerstemeier, Ajho, Alanpritt, Alienturnedhuman, Alongseptember, Andre Engels, Andrewleeder, Angmering, Angr, AnyFile, Archanamiya, Bblfan, Beland, Bender235, Bentogoa, Bigrich, Blanchardb, Blehfu, Bmcln1, Bob1960evens, Bobo192, Boleslavica, Bonalaw, BramvR, BrownHairedGirl, CU4ever, Calmer Waters, Can't sleep, clown will eat me, Charles Matthews, Charlesdrakew, Chris j wood, ChrisTheDude, Ckatz, ClamDip, ClickRick, Cliff Topp, Cnyborg, Coemgenus, Collard, Costesseyboy, DGG, Dah31, Dale Arnett, Dallan72, Dana boomer, Danielgibbins, Darklilac, Darkninja98, David Lion-West, Daytona2, Deb, Deflective, DevDevDevon, Discospinster, DitzyNizzy, DiverScout, DrKiernan, EH74DK, Edward, Ehrenkater, Epbr123, Erianna, EricITOworld, Euryalus, Excirial, Farosdaughter, GagHalfrunt, Ganymead, Ghoffman, Givsav, GoingBatty, Gongshow, Gonzo Baggins, Greenpenwriter, Guerongi, Gurchzilla, Hakluyt bean, Hammer1980, Hardyplants, Heron, Hmains, Hugo999, Humansdorpie, Igor22121976, Istabo, J.delanoy, JH49S, JPMcGrath, JaGa, Jake-helliwell, Jam1313, Jehricou99, Jeni, Jer ome, Jim1138, JustAGal, Jza84, Kbdank71, Khukri, KingOfTheLynn, Kinitawowi, Kresspahl, Kwamikagami, Lachramancy, Latricoteuse, Leahtwosaints, Les woodland, Lewisskinner, Lightmouse, LilHelpa, Limideen, Lingotic, Lord Cornwallis, Lugnuts, Lupin, MJCdetroit, MangoMango, Mareve, Martinslater, Marudubshinki, Masterknighted, Mathiasrex, Matt Crypto, McCrickerdXZY, Mellery, Mervyn, Mhockey, Michael Dorosh, Mintrick, Mndkingslynn, Moremadness2001, Morwen, Mutual Friend, Narson, Neddyseagoon, Nick Levine, Ninetyone, Nk, Nono64, Norfolkdumpling, Norm, North Wolf Inuit, Norwikian, OP41, Ondundozonananandana, One Night In Hackney, Oneblackline, Orioane, Our Phellap, OwenBlacker, Owlsandbadgersperformingliveforyouatcarnegiehall, Oxyman42, Palmiped, Paulajhaw, Pearle, Peridon, Phoe, Pinacolada1066, Pistachio disguisey, Pit-yacker, Plucas58, Pontificalibus, Ranveig, Ratarsed, Ravinpa, Rawclaw, Redvoler, Regan123, Renata, Richard Harvey, RobertWalden, Robertgreer, Ronhjones, Samthemaniloveut, Sandybag, Scott Bywater, Shelfsidestew, Shimane22, Shrinkness, Sigma714, SimonD, Simple Bob, Sir Stanley, Squids and Chips, Starbois, Stavros1, Stegbeetle, Stephenb, Stevieboyiv, Struway2, TGC55, TYelliot, Taurusk, TenPoundHammer, The Rambling Man, The Thing That Should Not Be, Thecheesykid, Tpbradbury, Tricky Victoria, Trusilver, Tutmosis, Twice25, Velella, Vinoir, Warofdreams, Wayne Slam, Wereon, WhaleyTim, Whitlam Warriors, Whpq, Wikiklrsc, Wimstead, WiseCat, Woohookitty, Wysprgr2005, Xerxes314, Xn4, Xxxeditxxx, Xymmax, Zik2, Zionandbabylon, 405 anonymous edits

Great_Yarmouth *Source*: http://en.wikipedia.org/w/index.php?title=Great_Yarmouth *Contributors*: 06readc, 21stCenturyGreenstuff, Acalamari, Adam Carr, Aj4444, Alethe, Alienturnedhuman, Amigadave, Anubis99, Arb, Arobson, Arsenalchampsleague, Ashley.lucas, Asterion, AxG, Baa, BarretB, Bbb2007, Bedesboy, Bellhalla, Benandorsqueaks, Billinghurst, Blatchfordm, Bloater44, Blue Square Thing, Bobo192, Britmax, Btline, BuddyHoliday, Bunnyhop11, C1self, CLW, Canterbury Tail, Caulde, Celardore, Chaosdruid, Charles Matthews, Charlesdrakew, Chasingsol, Chaz1dave, Chenzw, Chris j wood, Chrisdoyleorwell, Clear air turbulence, ClickRick, Cnyborg, Coralshin, Costesseyboy, Cubalupa, Custardninja, DRichardinson, DabMachine, Dace83, Danbooi, Danjamz, Darkcover21, De728631, Deb, Delpino, Demetrioudude, DerHexer, Detailman998, Dmgdj, Donotstandup, Dvavasour, Ed251191, Edward, Eenu, Efficacy, Elrith, Epbr123, Epistemos, Erpham, Felix Folio Secundus, Finavon, Flauto Dolce, Fluoronaut, FruitMonkey, Gantlion, Ghewgill, Gingerblokey, Gobold, Gr8opinionater, GraemeLeggett, GreenmanTGM, Gringotsgoblin, Harasseddad, HarryREDUX, Haydnaston, HelioSmith, Ian Cheese, Ian1000, IdreamofJeanie, Ilikeeatingwaffles, Innotata, Iridescent, J heisenberg, Jakeydude10, Jay-Capstaff, Jeremy Bolwell, Jllm06, John FitzGerald, Johnteslade, Jonbirduk, Juggertrout, Jvhertum, Kaj1mada, Karl1587, KayG, Keith Edkins, Klilidiplomus, Kresspahl, League Octopus, Les woodland, Lifebaka, Lightmouse, LilHelpa, LindsayH, Llywrch, Lolatyou94, Lord Cornwallis, Lukenewman, Lupin, Marek69, Mark Wheaver, Matt Crypto, Mellery, Mervyn, Militantblackguy2008, Mjroots, Mnnnbbvvccxxzz, Mnpeter, Mogador1653, Morwen, Neddyseagoon, NeilN, NigelR, NinetyCharacters, Nono64, Norfolkadam, Norm, Normal4norfolk, Northmetpit, Oakington, Ohconfucius, Once in a Blue Moon, Oneblackline, Orioane, Our Phellap, OwenBlacker, Palmiped, Paul Barlow, Pen of bushido, Pestridge, Philbarker, Philip Trueman, Pi.1415926535, Pit-yacker, Pol098, RHaworth, RJASE1, RMHED, Rambo's Revenge, Ratarsed, Regan123, Renata, Rich Farmbrough, Richard Harvey, Ridersden, Rjwilmsi, Roaringboy, Rob122, Robertgreer, Rror, Sc147, Scottfriz, Sd31415, Shadygrove2007, Sharpdarts, ShelfSkewed, SiJa2007, Silverhelm, Small-town hero, Smithtom1986, Snowmanradio, Stavros1, Stephen, Stephenb, Straightouttacomptonboy, Synchronism, TAS, The Rambling Man, Themfromspace, Thingg, Thumperward, Thurls, Tim Brook, Tourism Team, Twaz, Twiceuponatime, Ulric1313, Utcursch, Velella, Versus22, Victuallers, Vinoir, Vrenator, Vsmith, Warofdreams, Wereon, Wham2001, Wimt, WoodstockBraaap, Woohookitty, Wrightyboy, Xn4, Yarco, Zidonuke, Zoicon5, 278 anonymous edits

Central_London *Source*: http://en.wikipedia.org/w/index.php?title=Central_London *Contributors*: A. B., Achangeisasgoodasa, Afshin1, Ahoerstemeier, Alai, Andy, Andycjp, Anskas, BaldBoris, Barryob, Bevo74, Billinghurst, BrainyBabe, Bydand, CS46, CalJW, Calsicol, Can't sleep, clown will eat me, CarolGray, Chicheley, Colonies Chris, Coolhawks88, Djr xi, Ed g2s, Edward, Egil, Fifty Percent Normal, Gurch, Henrygb, Hmains, Hopex, Idamhar ila, Jbelien, Johndarrington, Josh Parris, Kbthompson, Ken Gallager, Kpflude, Lear's Fool, Lee setters, Lfh, Lightmouse, LilHelpa, Londonisms, Lozleader, Luna Santin, MRSC, Maurice45, Maxf, Mediaexpert, Melathron, Michael Hardy, Morwen, MykReeve, Natl1, Nev1, Nickfraser, Nomad2u001, Onebravemonkey, Otebig, Pcpcpc, PeterHuntington, Pointillist, Psemmusa, Raptornet, RexNL, Richard75, Richhoncho, SE7, SQL, Saddhiyama, SandyDancer, Shcha, Sladegreenforum, Stevertigo, Tagishsimon, Tankerlator, Tanthalas39, Tarquin Binary, The Giant Puffin, The Rationalist, Tim211010, Tom-, Triquetra, WikiWitch, Ywmpq205, 61 anonymous edits

Dereham *Source*: http://en.wikipedia.org/w/index.php?title=Dereham *Contributors*: Acalamari, Acenditcantu, Acrumpton, ApprenticeFan, Babylon77, Baranfin, Bearcat, Charlesdrakew, Chrisdoyleorwell, ClickRick, Costesseyboy, Cpl Syx, Dace83, Delta 51, DiverScout, Edwardx, Ehrenkater, Evud, Felix Folio Secundus, GraemeLeggett, H2005uk, Hadrianheugh, JaGa, Jesusisawsome, Jevansen, Jkncon, Johnbaud32, Kayau, LWF, Llsmartguy, Lozleader, MRSC, Mafsteele, Mjroots, Motmit, Nellyhench, Niceguyjoey, Norm, Northmetpit, Oneblackline, Palmiped, Pebbens, Pit-yacker, Plasticspork, Plastikspork, RHaworth, Rich Farmbrough, Richardguk, Sa2328, Shas, Squids and Chips, Stavros1, StevieNic, Tedmund, TheChaoticLord, Thomato, Tide rolls, Tommy2010, Warofdreams, Wetman, William Avery, Woohookitty, 73 anonymous edits

Swaffham *Source*: http://en.wikipedia.org/w/index.php?title=Swaffham *Contributors*: ABMx, Acalamari, Aggrav8r, Amakuru, Anthony Appleyard, Babylon77, Bcfoasis, Bearcat, Blackadderajr, Brad, Can't sleep, clown will eat me, Capricorn42, Charlesdrakew, Chris j wood, Chris the speller, Chuunen Baka, ClickRick, Czar Brodie, De728631, Dearagon, Debnigo, Edwardx, Epbr123, Finlay McWalter, Franz-kafka, Gaius Cornelius, Gdr, Geni, Grutness, Hadrianheugh, Hejordan, Hmains, I Hate CAPTCHAS, Inwind, JoanneB, JoeBlogsDord, Ligulem, Lord Cornwallis, Lozleader, Lumos3, Lupin, M.dibbly, MRSC, Mahatmacanejeeves, Mellery, MikeHobday, MrsHenry, Nedrutland, NeilN, Norm, Northmetpit, Ombt, Oneblackline, Palmiped, Pharillon, Pit-yacker, Professor marginalia, RedHillian, Reedjr5746, Regan123, Renata, Rich Farmbrough, Richard D. LeCour, Sigurd Dragon Slayer, Sloman, StAnselm, Starbois, Stavros1, Tabletop, Tardisrosedalek, Teratornis, Trog23, Vinoir, Warofdreams, Wcwtlc94, Wereon, Xxxeditxxx, 69 anonymous edits

Fransham *Source*: http://en.wikipedia.org/w/index.php?title=Fransham *Contributors*: Bearcat, ClickRick, Jeni, Nono64, Reedjr5746, Stavros1

Image Sources, Licenses and Contributors

File:All Saints Church - geograph.org.uk - 1263810.jpg *Source*: http://en.wikipedia.org/w/index.php?title=File:All_Saints_Church_-_geograph.org.uk_-_1263810.jpg *License*: unknown *Contributors*: Evelyn Simak

file:Norfolk UK location map.svg *Source*: http://en.wikipedia.org/w/index.php?title=File:Norfolk_UK_location_map.svg *License*: unknown *Contributors*: User:Nilfanion

File:Red pog.svg *Source*: http://en.wikipedia.org/w/index.php?title=File:Red_pog.svg *License*: unknown *Contributors*: Anomie

File:County Flag of Norfolk.svg *Source*: http://en.wikipedia.org/w/index.php?title=File:County_Flag_of_Norfolk.svg *License*: unknown *Contributors*: Arms_of_Norfolk.svg:

File:Norfolk UK locator map 2010.svg *Source*: http://en.wikipedia.org/w/index.php?title=File:Norfolk_UK_locator_map_2010.svg *License*: unknown *Contributors*: User:Nilfanion

Image:Arms of Norfolk.svg *Source*: http://en.wikipedia.org/w/index.php?title=File:Arms_of_Norfolk.svg *License*: unknown *Contributors*: BrightRaven, Bvs-aca, Darwinius, Fvasconcellos, Jza84, Mattes

Image:NorfolkNumbered.png *Source*: http://en.wikipedia.org/w/index.php?title=File:NorfolkNumbered.png *License*: unknown *Contributors*: Kanguole, Michiel1972

File:Loudspeaker.svg *Source*: http://en.wikipedia.org/w/index.php?title=File:Loudspeaker.svg *License*: unknown *Contributors*: Bayo, Gmaxwell, Husky, Iamunknown, Mirithing, Myself488, Nethac DIU, Omegatron, Rocket000, The Evil IP address, Wouterhagens, 18 anonymous edits

Image:Wells-next-the-Sea 1.jpg *Source*: http://en.wikipedia.org/w/index.php?title=File:Wells-next-the-Sea_1.jpg *License*: unknown *Contributors*: User:chris_j_wood

Image:The Wensum under trees.JPG *Source*: http://en.wikipedia.org/w/index.php?title=File:The_Wensum_under_trees.JPG *License*: unknown *Contributors*: User:Angmering

Image:NorwichCathedralSpire.JPG *Source*: http://en.wikipedia.org/w/index.php?title=File:NorwichCathedralSpire.JPG *License*: unknown *Contributors*: Original uploader was J-wiki at en.wikipedia

Image:NorwichRCC.JPG *Source*: http://en.wikipedia.org/w/index.php?title=File:NorwichRCC.JPG *License*: unknown *Contributors*: User:Angmering

Image:ElmHill.jpg *Source*: http://en.wikipedia.org/w/index.php?title=File:ElmHill.jpg *License*: unknown *Contributors*: Original uploader was TourNorfolk at en.wikipedia

Image:Mundesleybeachnorth.jpg *Source*: http://en.wikipedia.org/w/index.php?title=File:Mundesleybeachnorth.jpg *License*: unknown *Contributors*: Original uploader was TourNorfolk at en.wikipedia

Image:WroxhamBridge.jpg *Source*: http://en.wikipedia.org/w/index.php?title=File:WroxhamBridge.jpg *License*: unknown *Contributors*: TourNorfolk

Image:Peddars Way - Holkham Bay.jpg *Source*: http://en.wikipedia.org/w/index.php?title=File:Peddars_Way_-_Holkham_Bay.jpg *License*: unknown *Contributors*: User:Michael Perryman

Image:Britpieryarmouth.JPG *Source*: http://en.wikipedia.org/w/index.php?title=File:Britpieryarmouth.JPG *License*: unknown *Contributors*: Adam Smith. Original uploader was Gingerblokey at en.wikipedia

Image:Norwich Theatre Royal.JPG *Source*: http://en.wikipedia.org/w/index.php?title=File:Norwich_Theatre_Royal.JPG *License*: unknown *Contributors*: Northmetpit

Image:NorwichPlayhouse.JPG *Source*: http://en.wikipedia.org/w/index.php?title=File:NorwichPlayhouse.JPG *License*: unknown *Contributors*: C1self

Image:King's Lynn collage.png *Source*: http://en.wikipedia.org/w/index.php?title=File:King's_Lynn_collage.png *License*: unknown *Contributors*: User:Scott Bywater

File:Hanseatic Warehouse King's Lynn.JPG *Source*: http://en.wikipedia.org/w/index.php?title=File:Hanseatic_Warehouse_King's_Lynn.JPG *License*: unknown *Contributors*: Alienturnedhuman (talk). Original uploader was Alienturnedhuman at en.wikipedia

Image:Kings-lynn-customs-house.JPG *Source*: http://en.wikipedia.org/w/index.php?title=File:Kings-lynn-customs-house.JPG *License*: unknown *Contributors*: Original uploader was Alienturnedhuman at en.wikipedia

Image:Kingslynnfrontage.jpg *Source*: http://en.wikipedia.org/w/index.php?title=File:Kingslynnfrontage.jpg *License*: unknown *Contributors*: Bjh21, Collard, Kurpfalzbilder.de, Optimist on the run, Para, Ratarsed, Thryduulf, 1 anonymous edits

Image:Kings-lynn-river-great-ouse.JPG *Source*: http://en.wikipedia.org/w/index.php?title=File:Kings-lynn-river-great-ouse.JPG *License*: unknown *Contributors*: Original uploader was Alienturnedhuman at en.wikipedia

Image:Kings-lynn-campbells-soup-tower.JPG *Source*: http://en.wikipedia.org/w/index.php?title=File:Kings-lynn-campbells-soup-tower.JPG *License*: unknown *Contributors*: The original photograph was taken by myself (Ben Dickson) on November 28th 2006 with a Sony HDR-HC3 digital camcorder. Original uploader was Alienturnedhuman at en.wikipedia

Image:King's Lynn & West Norfolk.jpg *Source*: http://en.wikipedia.org/w/index.php?title=File:King's_Lynn_&_West_Norfolk.jpg *License*: unknown *Contributors*: Calmer Waters, KingOfTheLynn, Sreejithk2000, Theleftorium

Image:King's Lynn badge.jpeg *Source*: http://en.wikipedia.org/w/index.php?title=File:King's_Lynn_badge.jpeg *License*: unknown *Contributors*: KingOfTheLynn, Melesse

Image:RIVER GAYWOOD the mouth of the river at Kings Lynn joining the river Great Ouse 12th March 2007.JPG *Source*: http://en.wikipedia.org/w/index.php?title=File:RIVER_GAYWOOD_the_mouth_of_the_river_at_Kings_Lynn_joining_the_river_Great_Ouse_12th_March_2007.JPG *License*: unknown *Contributors*: Original uploader was Stavros1 at en.wikipedia

File:Kings Lynn Modern Center.jpg *Source*: http://en.wikipedia.org/w/index.php?title=File:Kings_Lynn_Modern_Center.jpg *License*: unknown *Contributors*: User:Igor22121976

File:Magnify-clip.png *Source*: http://en.wikipedia.org/w/index.php?title=File:Magnify-clip.png *License*: unknown *Contributors*: User:Erasoft24

File:2009-02-02-vancouver-quarter-kings-lynn.jpg *Source*: http://en.wikipedia.org/w/index.php?title=File:2009-02-02-vancouver-quarter-kings-lynn.jpg *License*: unknown *Contributors*: Alienturnedhuman (talk). Original uploader was Alienturnedhuman at en.wikipedia

Image:Lynn routes.png *Source*: http://en.wikipedia.org/w/index.php?title=File:Lynn_routes.png *License*: unknown *Contributors*: User:EricITOworld

Image:365504 at King's Lynn.jpg *Source*: http://en.wikipedia.org/w/index.php?title=File:365504_at_King's_Lynn.jpg *License*: unknown *Contributors*: User:Collard

Image:King's Lynn Mart 1.jpg *Source*: http://en.wikipedia.org/w/index.php?title=File:King's_Lynn_Mart_1.jpg *License*: unknown *Contributors*: Alienturnedhuman (talk). Original uploader was Alienturnedhuman at en.wikipedia

Image:Great Yarmouth Town Hall.jpg *Source*: http://en.wikipedia.org/w/index.php?title=File:Great_Yarmouth_Town_Hall.jpg *License*: unknown *Contributors*: Adam Smith

File:Britannia Monument.jpg *Source*: http://en.wikipedia.org/w/index.php?title=File:Britannia_Monument.jpg *License*: unknown *Contributors*: Yarco

File:Greatyarmouth.jpg *Source*: http://en.wikipedia.org/w/index.php?title=File:Greatyarmouth.jpg *License*: unknown *Contributors*: User:Arobson

File:Greatyarmouthpanorama.JPG *Source*: http://en.wikipedia.org/w/index.php?title=File:Greatyarmouthpanorama.JPG *License*: unknown *Contributors*: User:Gingerblokey

File:Britpieryarmouth.JPG *Source*: http://en.wikipedia.org/w/index.php?title=File:Britpieryarmouth.JPG *License*: unknown *Contributors*: Adam Smith. Original uploader was Gingerblokey at en.wikipedia

File:Fishing boat YH 671 - geograph.org.uk - 1067992.jpg *Source*: http://en.wikipedia.org/w/index.php?title=File:Fishing_boat_YH_671_-_geograph.org.uk_-_1067992.jpg *License*: unknown *Contributors*: Ashley Dace

File:Great Yarmouth beach near the Winter Gardens.JPG *Source*: http://en.wikipedia.org/w/index.php?title=File:Great_Yarmouth_beach_near_the_Winter_Gardens.JPG *License*: unknown *Contributors*: Original uploader was Gingerblokey at en.wikipedia

File:Haven Bridge Lifted.jpg *Source*: http://en.wikipedia.org/w/index.php?title=File:Haven_Bridge_Lifted.jpg *License*: unknown *Contributors*: User:Norfolkadam

File:Aerial View of Great Yarmouth.jpg *Source*: http://en.wikipedia.org/w/index.php?title=File:Aerial_View_of_Great_Yarmouth.jpg *License*: unknown *Contributors*: User:Norfolkadam

File:Flag of France.svg *Source*: http://en.wikipedia.org/w/index.php?title=File:Flag_of_France.svg *License*: unknown *Contributors*: Anomie

File:Open street map central london.svg *Source*: http://en.wikipedia.org/w/index.php?title=File:Open_street_map_central_london.svg *License*: unknown *Contributors*: User:MRSC

Image:London_360°_Panorama_from_the_London_Eye.jpg *Source*: http://en.wikipedia.org/w/index.php?title=File:London_360°_Panorama_from_the_London_Eye.jpg *License*: unknown *Contributors*: User:Farwestern

File:Dereham market place.JPG *Source*: http://en.wikipedia.org/w/index.php?title=File:Dereham_market_place.JPG *License*: unknown *Contributors*: User:Charlesdrakew

Image:East Derehammap 1946.png *Source*: http://en.wikipedia.org/w/index.php?title=File:East_Derehammap_1946.png *License*: unknown *Contributors*: OS

Image:The Mid Norfolk railway, Dereham yard - geograph.org.uk - 1204450.jpg *Source*: http://en.wikipedia.org/w/index.php?title=File:The_Mid_Norfolk_railway,_Dereham_yard_-_geograph.org.uk_-_1204450.jpg *License*: unknown *Contributors*: Ashley Dace

File:East Dereham Mill.jpg *Source*: http://en.wikipedia.org/w/index.php?title=File:East_Dereham_Mill.jpg *License*: unknown *Contributors*: Roger Gilbertson

File:Cowper Church Sunday School, East Dereham - geograph.org.uk - 528937.jpg *Source*: http://en.wikipedia.org/w/index.php?title=File:Cowper_Church_Sunday_School,_East_Dereham_-_geograph.org.uk_-_528937.jpg *License*: unknown *Contributors*: Bill Sibley

Image:Swaffham map1946.png *Source*: http://en.wikipedia.org/w/index.php?title=File:Swaffham_map1946.png *License*: unknown *Contributors*: OS

File:Rotunda Swaffham.jpg *Source*: http://en.wikipedia.org/w/index.php?title=File:Rotunda_Swaffham.jpg *License*: unknown *Contributors*: John Salmon

Image:Swaffham.JPG *Source*: http://en.wikipedia.org/w/index.php?title=File:Swaffham.JPG *License*: unknown *Contributors*: Martyn Davies

Image:Swaffamkevinhale.jpg *Source*: http://en.wikipedia.org/w/index.php?title=File:Swaffamkevinhale.jpg *License*: unknown *Contributors*: Kevin Hale. Original uploader was Hadrianheugh at en.wikipedia

File:Canary and Linnet Little Fransham 23 06 2010.JPG *Source*: http://en.wikipedia.org/w/index.php?title=File:Canary_and_Linnet_Little_Fransham_23_06_2010.JPG *License*: unknown *Contributors*: stavros1

GNU Free Documentation License Version 1.2, November 2002 Copyright (C) 2000,2001,2002 Free Software Foundation, Inc. 59 Temple Place, Suite 330, Boston, MA 02111-1307 USA Everyone is permitted to copy and distribute verbatim copies of this license document, but changing it is not allowed.

0. PREAMBLE

The purpose of this License is to make a manual, textbook, or other functional and useful document "free" in the sense of freedom: to assure everyone the effective freedom to copy and redistribute it, with or without modifying it, either commercially or noncommercially. Secondarily, this License preserves for the author and publisher a way to get credit for their work, while not being considered responsible for modifications made by others. This License is a kind of "copyleft", which means that derivative works of the document must themselves be free in the same sense. It complements the GNU General Public License, which is a copyleft license designed for free software. We have designed this License in order to use it for manuals for free software, because free software needs free documentation: a free program should come with manuals providing the same freedoms that the software does. But this License is not limited to software manuals; it can be used for any textual work, regardless of subject matter or whether it is published as a printed book. We recommend this License principally for works whose purpose is instruction or reference.

1. APPLICABILITY AND DEFINITIONS

This License applies to any manual or other work, in any medium, that contains a notice placed by the copyright holder saying it can be distributed under the terms of this License. Such a notice grants a world-wide, royalty-free license, unlimited in duration, to use that work under the conditions stated herein. The "Document", below, refers to any such manual or work. Any member of the public is a licensee, and is addressed as "you". You accept the license if you copy, modify or distribute the work in a way requiring permission under copyright law. A "Modified Version" of the Document means any work containing the Document or a portion of it, either copied verbatim, or with modifications and/or translated into another language. A "Secondary Section" is a named appendix or a front-matter section of the Document that deals exclusively with the relationship of the publishers or authors of the Document to the Document's overall subject (or to related matters) and contains nothing that could fall directly within that overall subject. (Thus, if the Document is in part a textbook of mathematics, a Secondary Section may not explain any mathematics.) The relationship could be a matter of historical connection with the subject or with related matters, or of legal, commercial, philosophical, ethical or political position regarding them. The "Invariant Sections" are certain Secondary Sections whose titles are designated, as being those of Invariant Sections, in the notice that says that the Document is released under this License. If a section does not fit the above definition of Secondary then it is not allowed to be designated as Invariant. The Document may contain zero Invariant Sections. If the Document does not identify any Invariant Sections then there are none. The "Cover Texts" are certain short passages of text that are listed, as Front-Cover Texts or Back-Cover Texts, in the notice that says that the Document is released under this License. A Front-Cover Text may be at most 5 words, and a Back-Cover Text may be at most 25 words. A "Transparent" copy of the Document means a machine-readable copy, represented in a format whose specification is available to the general public, that is suitable for revising the document straightforwardly with generic text editors or (for images composed of pixels) generic paint programs or (for drawings) some widely available drawing editor, and that is suitable for input to text formatters or for automatic translation to a variety of formats suitable for input to text formatters. A copy made in an otherwise Transparent file format whose markup, or absence of markup, has been arranged to thwart or discourage subsequent modification by readers is not Transparent. An image format is not Transparent if used for any substantial amount of text. A copy that is not "Transparent" is called "Opaque". Examples of suitable formats for Transparent copies include plain ASCII without markup, Texinfo input format, LaTeX input format, SGML or XML using a publicly available DTD, and standard-conforming simple HTML, PostScript or PDF designed for human modification. Examples of transparent image formats include PNG, XCF and JPG. Opaque formats include proprietary formats that can be read and edited only by proprietary word processors, SGML or XML for which the DTD and/or processing tools are not generally available, and the machine-generated HTML, PostScript or PDF produced by some word processors for output purposes only. The "Title Page" means, for a printed book, the title page itself, plus such following pages as are needed to hold, legibly, the material this License requires to appear in the title page. For works in formats which do not have any title page as such, "Title Page" means the text near the most prominent appearance of the work's title, preceding the beginning of the body of the text. A section "Entitled XYZ" means a named subunit of the Document whose title either is precisely XYZ or contains XYZ in parentheses following text that translates XYZ in another language. (Here XYZ stands for a specific section name mentioned below, such as "Acknowledgements", "Dedications", "Endorsements", or "History".) To "Preserve the Title" of such a section when you modify the Document means that it remains a section "Entitled XYZ" according to this definition. The Document may include Warranty Disclaimers next to the notice which states that this License applies to the Document. These Warranty Disclaimers are considered to be included by reference in this License, but only as regards disclaiming warranties: any other implication that these Warranty Disclaimers may have is void and has no effect on the meaning of this License.

2. VERBATIM COPYING

You may copy and distribute the Document in any medium, either commercially or noncommercially, provided that this License, the copyright notices, and the license notice saying this License applies to the Document are reproduced in all copies, and that you add no other conditions whatsoever to those of this License. You may not use technical measures to obstruct or control the reading or further copying of the copies you make or distribute. However, you may accept compensation in exchange for copies. If you distribute a large enough number of copies you must also follow the conditions in section 3. You may also lend copies, under the same conditions stated above, and you may publicly display copies.

3. COPYING IN QUANTITY

If you publish printed copies (or copies in media that commonly have printed covers) of the Document, numbering more than 100, and the Document's license notice requires Cover Texts, you must enclose the copies in covers that carry, clearly and legibly, all these Cover Texts: Front-Cover Texts on the front cover, and Back-Cover Texts on the back cover. Both covers must also clearly and legibly identify you as the publisher of these copies. The front cover must present the full title with all words of the title equally prominent and visible. You may add other material on the covers in addition. Copying with changes limited to the covers, as long as they preserve the title of the Document and satisfy these conditions, can be treated as verbatim copying in other respects. If the required texts for either cover are too voluminous to fit legibly, you should put the first ones listed (as many as fit reasonably) on the actual cover, and continue the rest onto adjacent pages. If you publish or distribute Opaque copies of the Document numbering more than 100, you must either include a machine-readable Transparent copy along with each Opaque copy, or state in or with each Opaque copy a computer-network location from which the general network-using public has access to download using public-standard network protocols a complete Transparent copy of the Document, free of added material. If you use the latter option, you must take reasonably prudent steps, when you begin distribution of Opaque copies in quantity, to ensure that this Transparent copy will remain thus accessible at the stated location until at least one year after the last time you distribute an Opaque copy (directly or through your agents or retailers) of that edition to the public. It is requested, but not required, that you contact the authors of the Document well before redistributing any large number of copies, to give them a chance to provide you with an updated version of the Document.

4. MODIFICATIONS

You may copy and distribute a Modified Version of the Document under the conditions of sections 2 and 3 above, provided that you release the Modified Version under precisely this License, with the Modified Version filling the role of the Document, thus licensing distribution and modification of the Modified Version to whoever possesses a copy of it. In addition, you must do these things in the Modified Version: A. Use in the Title Page (and on the covers, if any) a title distinct from that of the Document, and from those of previous versions (which should, if there were any, be listed in the History section of the Document). You may use the same title as a previous version if the original publisher of that version gives permission. B. List on the Title Page, as authors, one or more persons or entities responsible for authorship of the modifications in the Modified Version, together with at least five of the principal authors of the Document (all of its principal authors, if it has fewer than five), unless they release you from this requirement. C. State on the Title page the name of the publisher of the Modified Version, as the publisher. D. Preserve all the copyright notices of the Document. E. Add an appropriate copyright notice for your modifications adjacent to the other copyright notices. F. Include, immediately after the copyright notices, a license notice giving the public permission to use the Modified Version under the terms of this License, in the form shown in the Addendum below. G. Preserve in that license notice the full lists of Invariant Sections and required Cover Texts given in the Document's license notice. H. Include an unaltered copy of this License. I. Preserve the section Entitled "History", Preserve its Title, and add to it an item stating at least the title, year, new authors, and publisher of the Modified Version as given on the Title Page. If there is no section Entitled "History" in the Document, create one stating the title, year, authors, and publisher of the Document as given on its Title Page, then add an item describing the Modified Version as stated in the previous sentence. J. Preserve the network location, if any, given in the Document for public access to a Transparent copy of the Document, and likewise the network locations given in the Document for previous versions it was based on. These may be placed in the "History" section. You may omit a network location for a work that was published at least four years before the Document itself, or if the original publisher of the version it refers to gives permission. K. For any section Entitled "Acknowledgements" or "Dedications", Preserve the Title of the section, and preserve in the section all the substance and tone of each of the contributor acknowledgements and/or dedications given therein. L. Preserve all the Invariant Sections of the Document, unaltered in their text and in their titles. Section numbers or the equivalent are not considered part of the section titles. M. Delete any section Entitled "Endorsements". Such a section may not be included in the Modified Version. N. Do not retitle any existing section to be Entitled "Endorsements" or to conflict in title with any Invariant Section. O. Preserve any Warranty Disclaimers. If the Modified Version includes new front-matter sections or appendices that qualify as Secondary Sections and contain no material copied from the Document, you may at your option designate some or all of these sections as invariant. To do this, add their titles to the list of Invariant Sections in the Modified Version's license notice. These titles must be distinct from any other section titles. You may add a section Entitled "Endorsements", provided it contains nothing but endorsements of your Modified Version by various parties--for example, statements of peer review or that the text has been approved by an organization as the authoritative definition of a standard. You may add a passage of up to five words as a Front-Cover Text, and a passage of up to 25 words as a Back-Cover Text, to the end of the list of Cover Texts in the Modified Version. Only one passage of Front-Cover Text and one of Back-Cover Text may be added by (or through arrangements made by) any one entity. If the Document already includes a cover text for the same cover, previously added by you or by arrangement made by the same entity you are acting on behalf of, you may not add another; but you may replace the old one, on explicit permission from the previous publisher that added the old one. The author(s) and publisher(s) of the Document do not by this License give permission to use their names for publicity for or to assert or imply endorsement of any Modified Version.

5. COMBINING DOCUMENTS

You may combine the Document with other documents released under this License, under the terms defined in section 4 above for modified versions, provided that you include in the combination all of the Invariant Sections of all of the original documents, unmodified, and list them all as Invariant Sections of your combined work in its license notice, and that you preserve all their Warranty Disclaimers. The combined work need only contain one copy of this License, and multiple identical Invariant Sections may be replaced with a single copy. If there are multiple Invariant Sections with the same name but different contents, make the title of each such section unique by adding at the end of it, in parentheses, the name of the original author or publisher of that section if known, or else a unique number. Make the same adjustment to the section titles in the list of Invariant Sections in the license notice of the combined work. In the combination, you must combine any sections Entitled "History" in the various original documents, forming one section Entitled "History"; likewise combine any sections Entitled "Acknowledgements", and any sections Entitled "Dedications". You must delete all sections Entitled "Endorsements".

6. COLLECTIONS OF DOCUMENTS

You may make a collection consisting of the Document and other documents released under this License, and replace the individual copies of this License in the various documents with a single copy that is included in the collection, provided that you follow the rules of this License for verbatim copying of each of the documents in all other respects. You may extract a single document from such a collection, and distribute it individually under this License, provided you insert a copy of this License into the extracted document, and follow this License in all other respects regarding verbatim copying of that document.

7. AGGREGATION WITH INDEPENDENT WORKS

A compilation of the Document or its derivatives with other separate and independent documents or works, in or on a volume of a storage or distribution medium, is called an "aggregate" if the copyright resulting from the compilation is not used to limit the legal rights of the compilation's users beyond what the individual works permit. When the Document is included in an aggregate, this License does not apply to the other works in the aggregate which are not themselves derivative works of the Document. If the Cover Text requirement of section 3 is applicable to these copies of the Document, then if the Document is less than one half of the entire aggregate, the Document's Cover Texts may be placed on covers that bracket the Document within the aggregate, or the electronic equivalent of covers if the Document is in electronic form. Otherwise they must appear on printed covers that bracket the whole aggregate.

8. TRANSLATION

Translation is considered a kind of modification, so you may distribute translations of the Document under the terms of section 4. Replacing Invariant Sections with translations requires special permission from their copyright holders, but you may include translations of some or all Invariant Sections in addition to the original versions of these Invariant Sections. You may include a translation of this License, and all the license notices in the Document, and any Warranty Disclaimers, provided that you also include the original English version of this License and the original versions of those notices and disclaimers. In case of a disagreement between the translation and the original version of this License or a notice or disclaimer, the original version will prevail. If a section in the Document is Entitled "Acknowledgements", "Dedications", or "History", the requirement (section 4) to Preserve its Title (section 1) will typically require changing the actual title.

9. TERMINATION

You may not copy, modify, sublicense, or distribute the Document except as expressly provided for under this License. Any other attempt to copy, modify, sublicense or distribute the Document is void, and will automatically terminate your rights under this License. However, parties who have received copies, or rights, from you under this License will not have their licenses terminated so long as such parties remain in full compliance.

10. FUTURE REVISIONS OF THIS LICENSE

The Free Software Foundation may publish new, revised versions of the GNU Free Documentation License from time to time. Such new versions will be similar in spirit to the present version, but may differ in detail to address new problems or concerns. See http://www.gnu.org/copyleft/. Each version of the License is given a distinguishing version number. If the Document specifies that a particular numbered version of this License "or any later version" applies to it, you have the option of following the terms and conditions either of that specified version or of any later version that has been published (not as a draft) by the Free Software Foundation. If the Document does not specify a version number of this License, you may choose any version ever published (not as a draft) by the Free Software Foundation. ADDENDUM: How to use this License for your documents To use this License in a document you have written, include a copy of the License in the document and put the following copyright and license notices just after the title page: Copyright (c) YEAR YOUR NAME. Permission is granted to copy, distribute and/or modify this document under the terms of the GNU Free Documentation License, Version 1.2 or any later version published by the Free Software Foundation; with no Invariant Sections, no Front-Cover Texts, and no Back-Cover Texts. A copy of the license is included in the section entitled "GNU Free Documentation License". If you have Invariant Sections, Front-Cover Texts and Back-Cover Texts, replace the "with...Texts." line with this: with the Invariant Sections being LIST THEIR TITLES, with the Front-Cover Texts being LIST, and with the Back-Cover Texts being LIST. If you have Invariant Sections without Cover Texts, or some other combination of the three, merge those two alternatives to suit the situation. If your document contains nontrivial examples of program code, we recommend releasing these examples in parallel under your choice of free software license, such as the GNU General Public License, to permit their use in free software.